The SOS Guide to Live Sound

Optimising Your Band's Live-Performance Audio

Paul White

NEW YORK AND LONDON

First published 2015
by Focal Press
70 Blanchard Road, Suite 402, Burlington, MA 01803

and by Focal Press
2 Park Square, Milton Park, Abingdon, Oxon OX14 4RN

Focal Press is an imprint of the Taylor & Francis Group, an informa business

Notices
Knowledge and best practice in this field are constantly changing. As new research and experience broaden our understanding, changes in research methods, professional practices, or medical treatment may become necessary.

Practitioners and researchers must always rely on their own experience and knowledge in evaluating and using any information, methods, compounds, or experiments described herein. In using such information or methods they should be mindful of their own safety and the safety of others, including parties for whom they have a professional responsibility.

Product or corporate names may be trademarks or registered trademarks, and are used only for identification and explanation without intent to infringe.

Library of Congress Cataloging in Publication Data
White, Paul, 1949– author.
The SOS guide to live sound: optimizing your band's live-performance audio/
Paul White.
pages cm
1. Rock groups – Instruction and study. 2. Sound – Recording and reproducing.
3. Rock concerts – Production and direction. I. Title.
MT733.7.W45 2014
781.49 – dc23
2013045205

ISBN: 978-0-415-84303-4 (pbk)
ISBN: 978-0-203-75812-0 (ebk)

Typeset in Helvetica Neue
by Florence Production Ltd, Stoodleigh, Devon, UK

Printed and bound in India by Replika Press Pvt. Ltd.

Table of Contents

Preface

Whilst there are some excellent books that delve deep into the theory of large-scale sound systems, I've always felt that many of these were a little too theoretical, and that what many people really need is practical guidance on getting a good sound when playing small venues, using their own back-line and PA system. In many ways, this is actually more challenging than working in large-scale venues, as the audience will always be hearing a blend of the backline and what comes over the PA, whereas in a stadium the contribution of the back-line is almost insignificant. Furthermore, the sound company setting up for a stadium gig can place the speakers where they'll give the best coverage, but a band playing in a typical pub or small club often has to settle on a compromise location for their speakers, which poses challenges both for coverage and for getting enough level before feedback sets in.

This book examines the issues you'll be faced with when working in smaller venues and covers not only the choice and operation of PA systems, but also how the backline can be organised so that it works with the PA system, rather than battling against it. Monitoring, wireless mics, digital versus analogue mixers, effects, automatic pitch correction, the use of computers on stage, cabling and DI strategies are all also covered. All these live-sound elements work in a symbiotic way, with each affecting the others, and adding up to the performance of the whole system. Even the role of that vital component, the mixing engineer, is examined in some detail.

Many of today's audience will have been raised on a diet of highly produced recorded music, and may therefore have high expectations of the sound at a live gig—inaudible vocals and overpowering, out-of-tune electric guitar will no longer impress! While the visual aspect of a live performance can help carry the show, even when the audio isn't quite up to studio standards, there's simply no excuse for badly-balanced sound, drowned vocals or tuning problems. Having spent much of my life either playing gigs or mixing them for other people, I feel the time is right to pass on all that I've learned over the years in the hope that some of those brave souls who support live music in smaller venues might be able to enjoy a better-sounding musical performance.

Paul White

Images

Paul White, Hugh Robjohns, Dave Lockwood, Mike Crofts, Chris Korff, George Nicholson-Hart

Chapter One
Drivers and Loudspeakers

Without loudspeakers there would be no PA systems and no guitar amplifiers, so any discussion of live sound really needs to kick off with a close look at these vital components, but I promise not to get too technical. Although there have been some departures from the most commonly employed 'moving-

The fundamental components of a moving-coil cone driver haven't changed since they were invented over a hundred years ago.

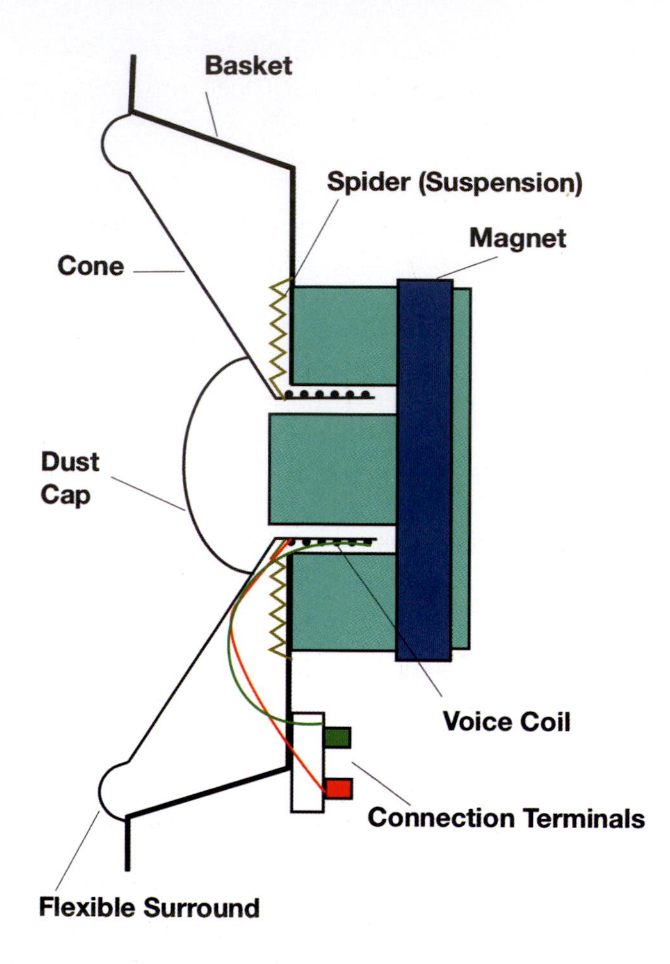

Loudspeaker cross-section.

coil' speaker principle, such as the electrostatic speakers that have enjoyed popularity in hi-fi circles, the overwhelming majority of live-sound loudspeaker systems use a moving-coil system that has changed little in principle over the best part of a century—although the designs have become much more efficient.

A loudspeaker without a cabinet is known as a 'driver' and pretty much everyone will be familiar with its general appearance, with a paper or composite-material cone, usually circular, suspended in a cast or pressed metal basket with a magnet fixed to the back. If you were to dismantle a typical driver you'd see that there's a coil of fine wire (the 'voice coil') wound onto an insulating cylinder (the 'former'), which is fixed to the back of the cone. The coil and former assembly projects rearwards into a narrow gap in the permanent magnet on the back of the basket, and is therefore sited within a strong magnetic field.

When an alternating current electrical signal is fed into the voice coil, it sets up a varying magnetic field that has the effect of pushing or pulling the voice coil relative to the fixed magnet in accordance with Faraday's Law and something called Fleming's Left Hand Rule. Don't worry if you've long-since forgotten all that stuff, as it certainly won't affect your ability to set up a good mix! As the voice coil is attached to the loudspeaker cone, its movement causes the latter to move back and forth in its flexible suspension, creating variations in sound pressure (sound waves) that correspond to the electrical input.

Modern loudspeaker drivers can handle far greater power and produce much higher sound-pressure levels (SPLs) than early models. This can be attributed both to progressive refinements of the fundamental design and to improved materials such as new magnetic alloys and adhesives that are able to withstand higher temperatures. This latter point is very important in the context of high-power audio systems, and I'll discuss it further later.

Precision engineering allows the magnetic gap to be made as narrow as possible so as to create a greater field strength, while distortion can be minimised by sophisticated suspensions, modifications to the shape of the magnetic pole-piece and by adding 'shorting rings' that shape the magnetic field to provide greater linearity over the travel of the voice coil. The development of sophisticated digital speaker-control systems that can monitor and regulate both average and peak power levels has also helped squeeze the maximum amount of level out of drivers without breaking them.

The further the cone of a driver can move while retaining a linear relationship with its electrical input, the greater the sound pressure level (SPL) it can produce before distortion sets in. The degree of linear excursion of which a driver is capable is given the term 'X-max' and this is one aspect of speaker design that has been improved over the years. As you might expect, a speaker cone can only move so far before it reaches its physical limit where either the suspension will permit no more travel or the voice coil will be forced out of the magnetic gap. When the cone excursion approaches these limits its physical movement is no longer proportional to the electrical input and so audible distortion occurs. The lower the audio frequency, the greater the cone excursion required to produce the same audible sound level so when a speaker is overdriven, it is usually the loud bass notes that push it over the edge into distortion, producing an unpleasant 'fuzzy' sound.

In order for the sound to have an acceptably low level of distortion, the cone must also hold its shape, which is not as easy as it may sound—in practice, the surface of any cone speaker will exhibit modal resonances known as breakup modes, much as a plucked guitar string produces harmonics as well as a fundamental pitch. By optimising the shape of the cone and using a material that offers the right combination of stiffness, self-damping and light weight, the audible effect of breakup modes can be minimised.

Even the best loudspeakers convert only a few percent of the applied amplifier power into sound, with the rest being transformed into heat. With some modern drivers being capable of handling 1000 watts or more, that means there's sometimes a lot of heat to dissipate! While high-temperature materials and adhesives have made a big contribution to increases in power handling, that unwanted heat still has to be channelled away from the voice coil via heat sinks and ventilation. Even during normal operation, the increased temperature of the voice coil and magnet system will reduce the loudspeaker's efficiency until it cools again—a phenomenon known as power compression. If more heat is being generated than can be dissipated, the voice-coil temperature will continue to rise and the coil will eventually burn out.

Dedicated high-frequency drivers, which we will take a look at shortly, sometimes use a special cooling liquid to fill the magnetic gap in which the voice coil is suspended, the most common being Ferrofluid. This comprises a colloidal suspension of magnetic material in a lubricating liquid that helps conduct heat away from the coil more quickly than relying on radiated conductance alone.

One reason why loudspeaker systems are so inefficient is that a moving cone of paper doesn't couple very effectively with the air it is trying to move—in mechanical terms there's an impedance mismatch, and this can be addressed with horn-loading techniques which are discussed later. In smaller speaker systems, of the type owned and transported by individual users, only the high frequency driver tends to be genuinely horn-loaded as the shorter wavelengths involved mean the horn flare can be made acceptably small.

The other limit on efficiency is attributable to the design of the driver itself: most drivers have an efficiency rating expressed as the sound pressure level (SPL) in decibels, measured at one metre, directly on axis with the centre of the cone, with the driver being fed with one watt of power, usually at a single frequency. This can't give the full picture as it doesn't take into account the polar pattern of the driver—i.e. how much energy is radiated off-axis—but it is a useful guide. Whilst typical hi-fi speakers may produce an SPL of as little as 85dB/W/m, it is not uncommon for PA drivers to be rated in excess of 100dB/W/m. Ultimately though, the efficiency of a PA speaker depends both on the driver itself and on the design of the cabinet it is placed in.

As high-powered amplifiers are now relatively cheap, many speaker designs sacrifice efficiency in favour of keeping the cabinet size small and then use 'brute force' amplification power

EFFICIENCY AND SPL

Efficiency and maximum SPL are not directly related and, similarly, the amount of power a speaker can tolerate tells you little about the maximum SPL it can produce. While the maximum quoted SPL is a useful figure to know, even that can still be misleading as these figures are often calculated from the driver's performance at lower power levels, rather than actually measured, and therefore take no account of any power compression caused by heating. So the actual maximum SPL during real-world usage may be a little less than the spec-sheet promises, especially if the average power level is high enough to keep the drivers running hot.

to achieve the necessary SPL—although this also increases the amount of heat that needs to be dissipated. As a general rule, it is not possible to combine high efficiency at low frequencies while maintaining a compact enclosure size, although some ingenious solutions have pushed the boundaries of what is possible.

Very high-powered amplifiers and sophisticated digital signal processing allow high SPLs to be generated from very compact enclosures, such as this Mackie DLM speaker.

Loudspeaker Power Ratings

Loudspeakers are usually rated on the assumption that the signal fed into them will comprise both loud peaks and some quieter sections. That's why you may see different power handling figures for music use, continuous sine-wave testing and short-term peaks. The use of compressor/limiters or significantly distorted sound sources increases the average signal level, which, at high SPLs, heats up the drivers and causes power compression, and as we've seen this can make the published loudspeaker power ratings and maximum SPL figure difficult to interpret.

Divide and Conquer

Building a single loudspeaker that can cover the entire frequency range of human hearing is a tough call as the requirements for creating high SPLs at low frequencies are very different from those required to reproduce high frequencies. Low frequencies have long wavelengths—100Hz has a wavelength of over three metres—and to get any kind of volume means moving a lot of air. Larger cones can obviously move more air for a given cone excursion which is why subwoofers usually incorporate speakers of 12-, 15- or 18-inch diameter, but these same speakers, which need to have very stiff cones, are of little use for reproducing high frequencies where the wavelengths involved are very short—the wavelength at 10kHz, for example, is just under 35mm. There are a number of reasons for large speakers being unsuitable for high frequency reproduction, the most obvious being that the inertia of a relatively heavy, large-diameter cone and the hefty voice coil stuck to the back of it, works against it accelerating and de-accelerating very quickly, as is required for high frequencies

TIP: As a general rule, speakers tolerate short periods of undistorted overload much better than prolonged periods of clipped input, so having some amplifier power in hand to allow a safety margin before clipping is always a good thing. Correctly set limiters can also help prevent clipping.

where the diaphragm changes direction thousands of times per second.

Another reason is to do with the physical width of the driver's cone, which at high frequencies may be several wavelengths across. At 1kHz the wavelength is just over one third of a metre—which happens to be around the same size as the diameter of a 12- or 15-inch cone driver. Each part of the cone generates sound by exciting the air molecules in front of it and if you're listening from directly in front of the speaker, the sound from different parts of the cone will arrive at pretty much the same time, although the very fact that it is conical and not flat means that even this generalisation can only be approximate. If the listening position is offset to one side, however, the sound from the edge of the cone closest to you reaches your ears slightly sooner than the sound from the furthest edge of the cone, and while the time difference may be under one

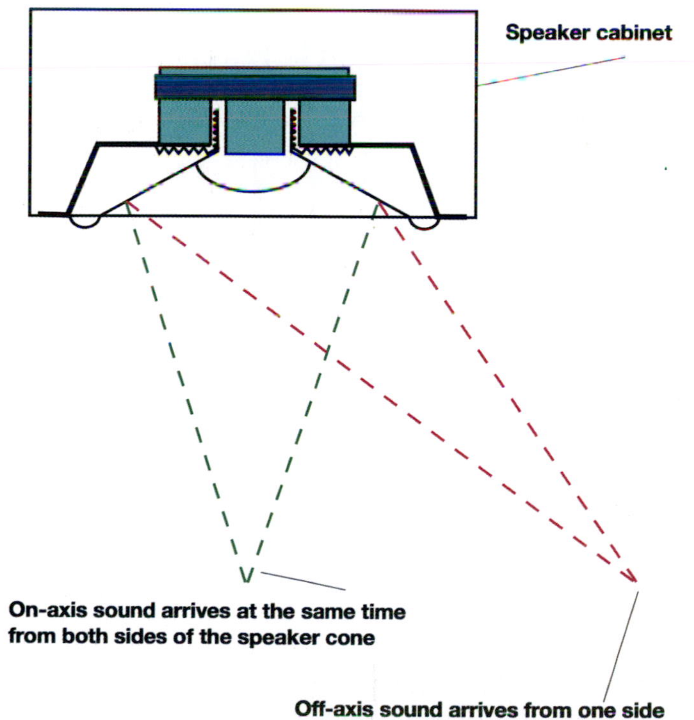

➤ Arrival time differences on- and off-axis cause audible anomalies in the frequency response.

millisecond, that's enough to cause phase cancellation at some frequencies and addition at others. These effects can become quite significant at frequencies above 1kHz or so.

Having different versions of the same sound arriving at slightly different times produces an audible effect we refer to as 'comb filtering', so-called because a graph of the frequency response looks somewhat like the teeth of a comb, with alternating peaks and dips in the response. These peaks and dips vary in position depending on frequency so when a complex sound, such as music, encounters a comb filter, the result is that we hear some frequencies being suppressed while others come over strongly. We tend to describe this sound quality as 'phasey' or coloured, and in subjective terms it's not unlike a mild version of the sound you'd hear from a flanger or phaser pedal with its sweep oscillator turned off.

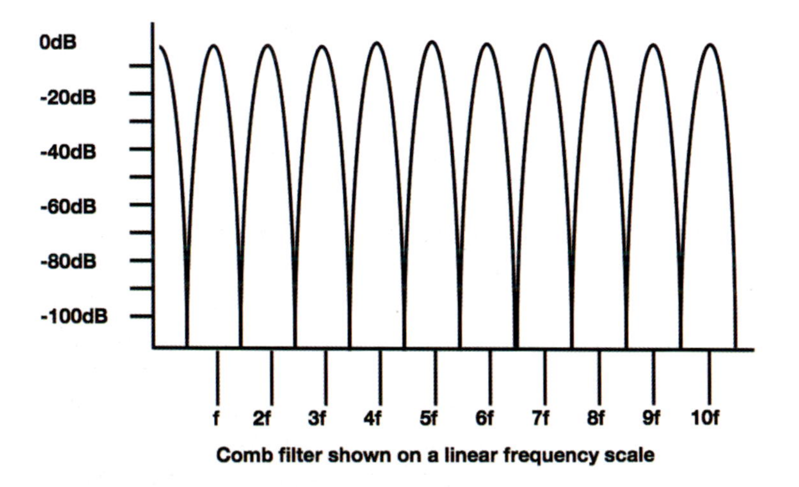

Response plot of comb filtering.

If you were to look at the polar plot of a large-diameter loudspeaker at different frequencies, you'd see a pattern of lobes rather than an even response, with most of the high-frequency energy being projected in a relatively tight beam along the axis of the speaker. This beam gets narrower as the frequency is increased, although the level of high frequencies eventually drops off as the inertia of the cone inhibits the high speed back-and-forth movement needed to reproduce high-frequency sound. As a ballpark figure, for example, a typical 12-inch driver might work happily up to 3kHz or thereabouts, but its efficiency will fall off quite quickly above that.

Polar plot of a 12-inch loudspeaker.

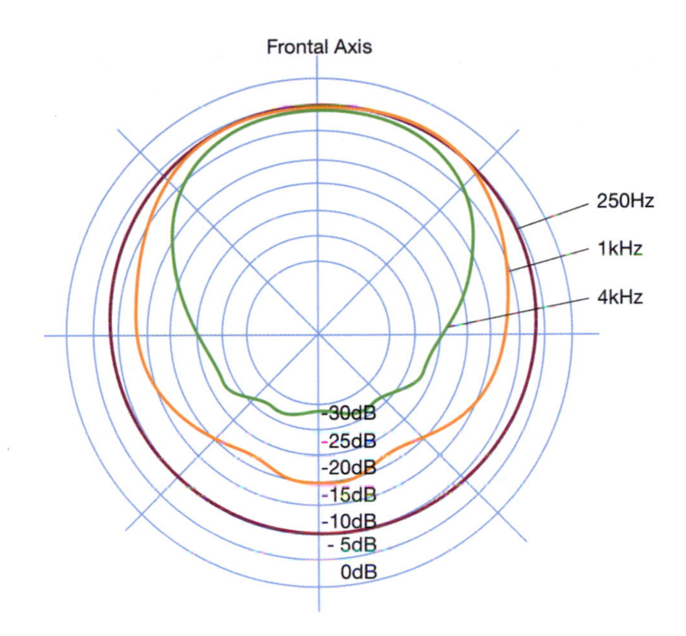

Bose pioneered the use of multiple small drivers in compact PA cabinets in the late 1970s.

A smaller diameter cone makes it easier to reproduce higher frequencies, but of course this reduces the effectiveness of the speaker when it comes to reproducing low frequencies at high SPLs, as it wouldn't be able to move so much air. Small speakers can reproduce low frequencies if they're loaded into the right kind of enclosure, but only at a much lower SPL. To get the same SPL as a larger diameter speaker, the small speakers have to be used in multiples. That's why some bass guitar cabinets come with four or even eight ten-inch speakers. There are also some PA systems that use multiple small cone drivers to cover the majority of the frequency range, although these are usually teamed with a subwoofer where deep bass is required. Bose were one of the first companies to build effective compact PA speakers using multiple small-diameter drivers (the Bose 802), though some electronic EQ had to be used to maintain a reasonably flat frequency response.

Practical Multi-way Systems

PA technology has actually changed quite a lot over the past 50 years: big systems have got much bigger, but the new technology also allows portable systems to be made smaller, lighter and more powerful, which is good news for gigging bands with limited transport capacity, or for those playing in small venues.

In the early days of touring bands, the PA system was likely to comprise a mono mixer amplifier of around 100 watts driving a couple of 4 × 10 or 4 × 12 speaker columns with no tweeters, no crossovers and no off-stage mixing. Effects were limited to maybe a spring reverb or tape-loop echo, and the microphones employed tended to be moving-coil models mostly by the likes of Shure or Beyer, or perhaps even fragile ribbon mics such as the ubiquitous Reslo. Basic concepts such as 'flat frequency response', 'controlled directivity' or even stage monitoring were understood only by the most visionary of audio pioneers, but fortunately much has changed since then, with the biggest advance being the use of multiple drivers to cover the audio-frequency spectrum with greater efficiency and fidelity.

Where high SPLs are needed, better performance can be obtained if the audio range is divided up between speakers of different sizes and types so that each can handle a frequency region for which it is optimised. This is achieved using a specially designed filter called a crossover. High-frequency-driver diaphragms are required to move much more rapidly than those of a bass driver and therefore have to be protected from low-frequency signals outside their operating range. A dedicated midrange speaker has both upper and lower limits of operation, so it has to be fed via both high- and low-pass filters to ensure that it receives only midrange frequencies. Similarly, bass drivers must be prevented from receiving frequencies higher than they are designed to reproduce, in order to prevent high-frequency beaming and a loss of sound quality.

Horns

A loudspeaker fixed into the face of a flat baffle board—a configuration called a 'direct radiator'—isn't a particularly efficient way of translating the speaker cone movement into air movement, especially at lower frequencies. Improving driver efficiency requires some kind of acoustic transformer or gearbox, which is where horn-loading comes in. A number of different horn-loaded bass speaker designs do exist, often with the horn 'folded' inside the cabinet to save space—though these may still be too large for the band hauling their own gear. They are more often employed in multiples for stadium concerts as part of a very large system. In smaller systems, lower efficiency is accepted as a trade-off against size.

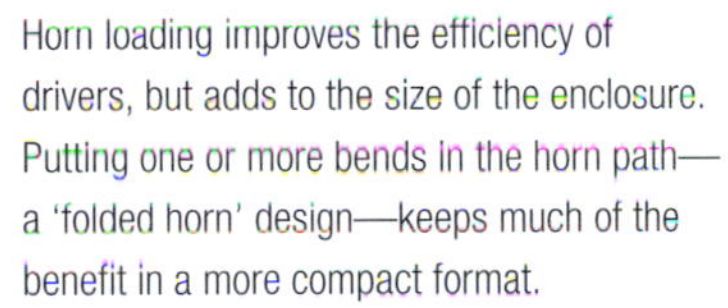
Horn loading improves the efficiency of drivers, but adds to the size of the enclosure. Putting one or more bends in the horn path—a 'folded horn' design—keeps much of the benefit in a more compact format.

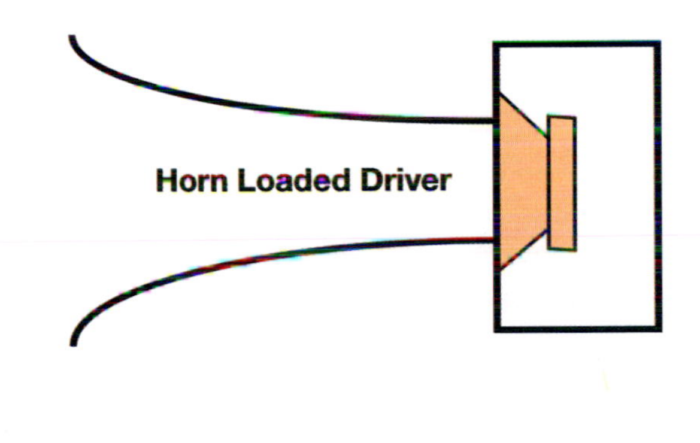

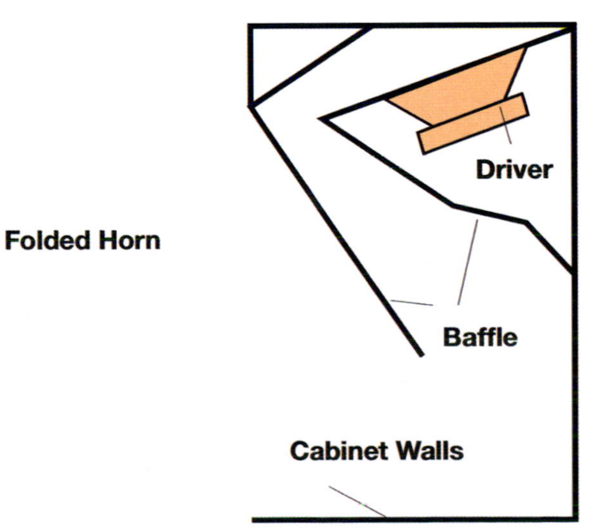

Midrange horns can be smaller because of the shorter wavelengths involved, but they still add significantly to the size of the enclosure. At one time, most pub-circuit bands could be seen struggling into small venues carting monstrous folded-horn bass-bin contraptions with them, but today's more efficient drivers and high-power, lightweight amplifiers make it possible to provide adequate volume without resorting to bulky horn-loaded cabinets. Therefore, as bass and mid horns are now far less relevant to portable PA systems I won't be examining their operation in any depth, and mention them here only for the sake of completeness.

High-frequency horns, on the other hand, are used in virtually all PA systems. For PA use, specialised dome-tweeter drivers known as compression drivers, used in conjunction with a horn-shaped 'flare' attached to the front, are very common even in compact 'plastic box' PA cabinets.

These horn flares have a narrow 'throat' that is smaller than the diameter of the compression driver dome and shaped to optimise the air pressure loading around the tweeter to increase its efficiency. The horn flare is shaped to control the directivity of the sound: a dome tweeter generates a spherical wavefront, so the angle over which the sound is spread is the same in both the horizontal and vertical planes, but for practical PA use, it is usually beneficial to limit the vertical dispersion and to widen the

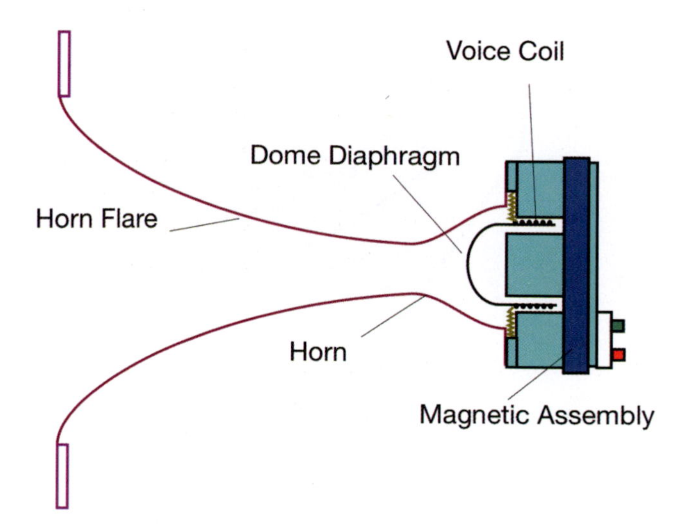

Horn loading both increases efficiency and controls the directivity of the driver.

horizontal dispersion, thereby concentrating more sound directly on the audience and wasting less on the floor and ceiling.

Some HF (high frequency) units are fitted with what is termed a 'constant directivity' horn whose profile differs slightly from that of the basic exponential flare. With an exponential-flare horn the dispersion angle narrows as the frequency increases. A constant directivity horn is shaped to make the dispersion more even across the HF range but requires a little built-in EQ (usually built into the crossover) to balance level discrepancies introduced by the changes made to the dispersion angle.

Understanding Crossovers

A crossover, as discussed above, is essentially a network of filters designed to send just a specific part of the audio spectrum to each driver, and depending on the size of the system, the spectrum will typically be divided into two or three separate frequency bands. The steepness of the filters is expressed in dB per octave and varies from one design to another. The simplest filter topology has a 6dB/octave slope and is said to be 'first order'. By chaining additional filter stages together the filter can be made steeper in 6dB/octave increments, giving 12dB/octave 'second order', 18dB/octave 'third order' or 24dB/octave 'fourth order' slopes. As well as ensuring that each driver covers only its designated frequency range, using steep-slope filters also reduces the area of overlap between drivers handling adjacent frequency bands. A wide overlap at the crossover point can lead to phase problems as both drivers deliver a slightly different version of the same audio signal in the overlap region. Low filter slopes also mean more 'out of band' signal reaching a driver and in the case of HF drivers in particular this can result in audible distortion or damage.

In smaller speakers, the crossover often comprises just simple 'passive' filtering circuitry built into the speaker cabinet, with the full-bandwidth output from the amplifier being split to send high frequencies only to the tweeter and low/mid frequencies only to the cone driver.

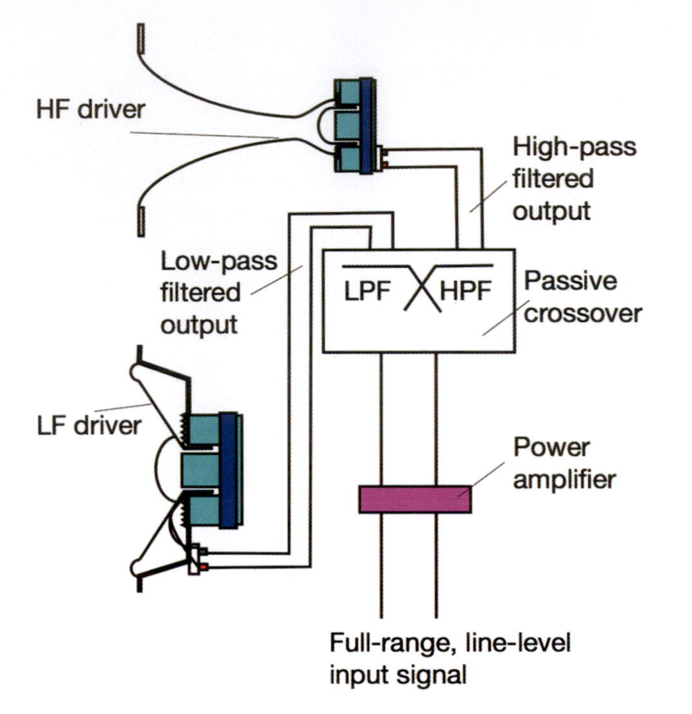

A passive two-way crossover, operating on a 'loudspeaker level' signal.

Passive designs are reliable and cost-effective, but are best suited only to relatively low-power systems, as they are inefficient, with some of the amplifier power being dissipated as heat within the passive crossover circuit. Furthermore, since the drivers in a multi-way system are rarely equally efficient, the more sensitive drivers have to be fed with attenuated signals to balance the overall response. This amounts to an almost 'deliberate' waste of power in order to achieve a flat frequency response!

Active Crossover Benefits

In an 'active' crossover system there's a separate power amplifier for each frequency range, and the filters are placed before the amplifiers so they only have to deal with line-level signals, rather than high-power amplifier outputs. System performance is no longer dictated by the driver with the lowest efficiency as the amplifier setup can be optimised to provide more power to less efficient speakers, which gives the system designers more flexibility in selecting drive units.

An active crossover works on the signal at line-level, before the power amplifiers.

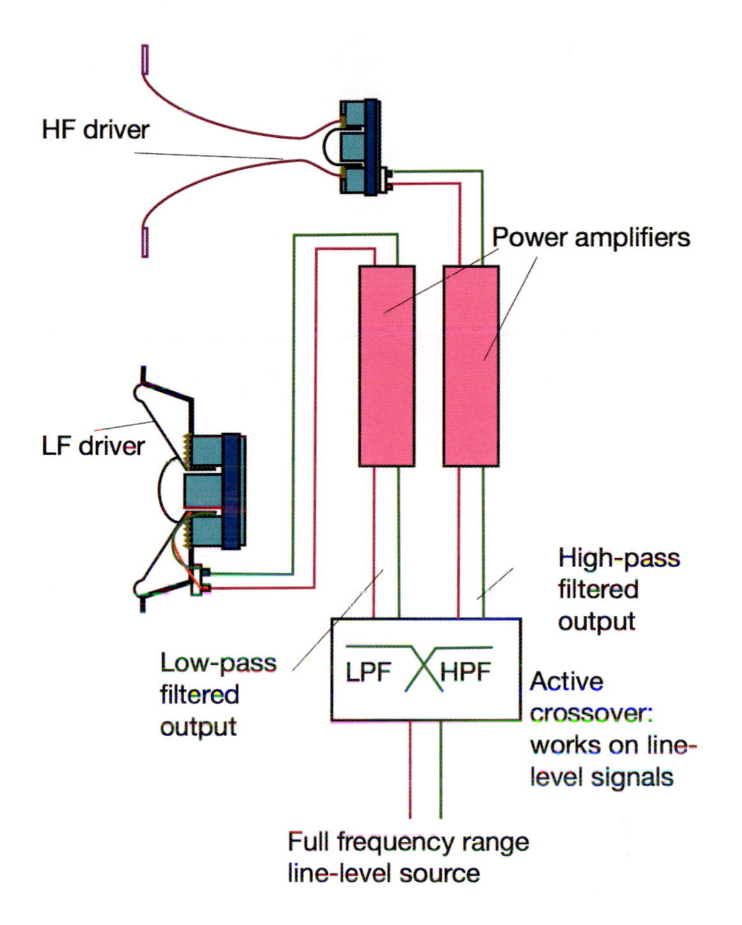

If you combine active crossovers with limiters, equalisers and delay circuitry—something that is relatively easy to do in the digital realm—you get what is referred to today as a 'speaker management system'. Effectively, you have everything you need to send the right signals to the right speakers at the right levels and at the right times, all in one box. Longer delays are included for use at large-scale events so that speakers in different locations can be time-aligned to allow for distance, or 'speed of sound' delay (approximately one foot per millisecond), but small-venue PA systems don't generally need to take this into consideration. However, some small timing adjustments may be made inside some active speaker systems to ensure that the sound from the high frequency and low frequency units are perfectly in phase.

The active crossovers built into some active subs are in fact more akin to full speaker management systems as they filter the signals reaching the bass speakers and also often provide a high-pass filtered line feed to the main speakers. They usually incorporate limiters to avoid amplifier clipping, too.

This latter point is very important as clipped audio contains a lot of high-frequency harmonics that, if fed into a tweeter, might destroy it with excess HF energy. If this doesn't seem intuitive, consider the following situation: a speaker incorporating a passive crossover is fed from an amplifier that has been driven into clipping by being fed too high an input signal. Each time a loud bass note or drum beat is played, the amplifier clips, producing square waves, rich in high-frequency, odd-order harmonics. These feed through the passive crossover in exactly the same way as legitimate high-frequency signals, so they reach the tweeter. If these artificially generated harmonics are high enough in level and long enough in duration, they will cause the voice coil to overheat and burn out.

Most cabinets with two-way crossovers will only extend down to somewhere between 80 and 45Hz, depending on their size and design, so to fill in the lows below this (essential if you're amplifying bass guitar and kick drum to any significant degree) a subwoofer is often added, which might typically reinforce or take over from the main speakers at somewhere between 120 to 80Hz and going down as low as 30Hz, effectively creating a three-way system.

Enclosures

The box that houses the drivers is much more than just a box. When a loudspeaker cone moves forward it compresses the air directly in front of it, and at the same time the pressure of the air behind the cone is lowered. Conversely, when the cone moves backwards, the air behind is compressed while the pressure directly in front of the cone is reduced. This is how the driver sets up sound waves in the air. However, if the driver is in free space, rather than in a box of some kind, the high-pressure air simply escapes around the edge of the driver into the low pressure area and restores equilibrium, so the majority of the energy is wasted pumping air from the front of the driver to the back rather than creating sound.

To prevent the air leaking around the sides of the driver, it can be mounted onto a large baffle board, and with an infinitely large and solid baffle, we'll only hear the sound generated in front of the driver. Obviously a huge baffle isn't practical for portable sound systems, but what if we fold our baffle into the shape of a closed box to trap all the sound from the rear of the speaker? Maybe we can also put some acoustic absorbers in the box to prevent reflected sound interfering with the movement of the speaker cone? For obvious reasons, this type of box is known as a 'Sealed Cabinet' enclosure (sometimes incorrectly referred to as an 'Infinite Baffle'—a true infinite baffle doesn't load the driver in the same way) but as it is impossible to absorb all the energy coming from the rear of the speaker cone, the box needs to be quite large in order to reproduce low frequencies effectively. Sealed Cabinet designs are commonly used for hi-fi, studio monitoring and small satellite speakers that don't need to reproduce low frequencies at very high SPLs, but what if that wasted energy from the rear of the speaker could be put to better use?

Bass Ports

Portable PA systems are generally designed to generate as high an SPL as possible from the smallest practical enclosure, and whilst choosing an efficient driver is a good start, it is very common to use what is known as a 'ported' or 'Bass Reflex' enclosure to improve the low-frequency performance. This isn't as effective as a garage-door sized horn flare, but is far more compact.

If you make a hole in a sealed-box enclosure and put a short tube through the hole, the box becomes a resonator that, when 'excited' by the sound coming from the rear of the speaker cone, produces a fixed tone, rather like blowing over the mouth of a bottle. The frequency and 'Q' (bandwidth) of the resulting resonance depends on the volume of air inside the box, the characteristics of any of physical damping material, the dimensions of the port (diameter and length) and the mechanical 'springiness' of the loudspeaker. If the Q is made too high, the bass will tend to 'boom' at a single frequency, but done correctly with the enclosure resonance tuned to be just below the point where the bass response of the speaker starts to roll off, it can extend the low-frequency response of a speaker to a useful degree.

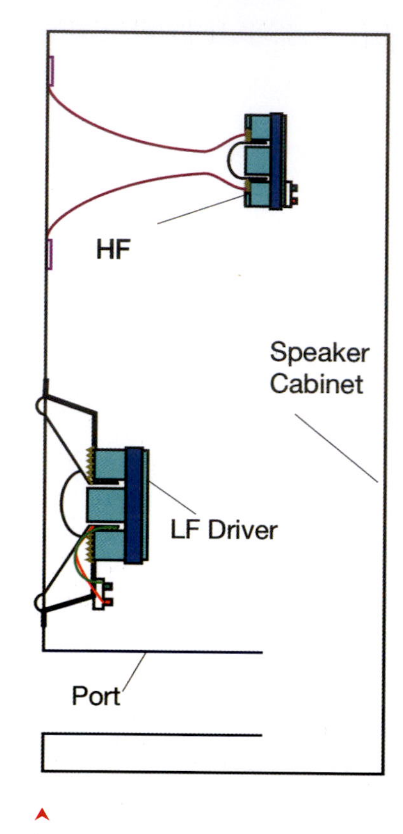

A tuned port can extend the bass response of a driver and cabinet beyond what might be achieved with a sealed box.

Once a port has been fitted to a cabinet, there's no natural 'air spring' inside the box to cushion the rear of the driver at very low frequencies, so it is then important to filter out frequencies that are below the resonant frequency of the port in order to prevent the driver from trying to move too far and damaging itself. Full-range, portable PA cabinets containing HF and LF drivers are usually ported, as are many subwoofers. You can't get away from the general rule that larger diameter drivers tend to require larger cabinets to work properly but at least they can now be made to fit into a car rather than a large van.

The most practical subwoofers for smaller-venue use tend to be active, ported designs and include the necessary crossovers to split the incoming signal between the subwoofer itself and the main speaker system. A single sub may be used in an otherwise stereo system as most of the directional cues come from the higher frequencies—the fact that the subwoofer itself is mono is not really noticeable. Typical subs for use with 10- or 12-inch full-range main speakers are powered by 15-inch or 18-inch drivers, although there are smaller systems that use 12-inch or 2 × 10-inch subs.

Band-pass Subwoofer Enclosures

In a band-pass subwoofer enclosure, the speaker is mounted on an internal baffle with a tuned, ported cavity in front of it and another enclosure behind it that may or may not be ported. This is an efficient system that also reduces the amount of cone movement necessary to produce the required SPL, although thermal driver stress may still occur. The practical implementation may also be around 50 percent larger than a traditional ported enclosure. As a rule, the more

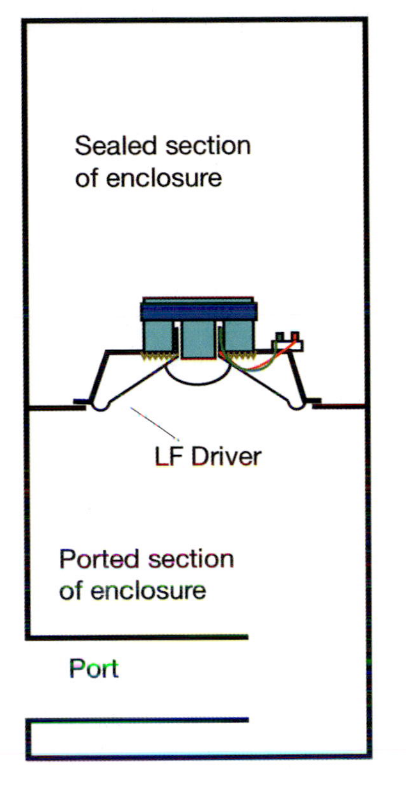

A band-pass enclosure configuration, favoured for sub-woofers.

a bass cabinet relies on cabinet tuning to increase efficiency, the more its bass notes will exhibit 'overhang' or extension. This can make the bass notes seem louder, but often at the expense of 'tightness'.

SUBWOOFER PLACEMENT

Subwoofers work most efficiently when sited directly on the floor, as positioning them above the ground on staging or risers runs the risk of some frequencies reflecting from the ground causing cancellation, resulting in some bass notes being louder than others. You'll also get more SPL from a subwoofer if it is set up close to a wall, due to what is known as 'boundary effect'—at very low frequencies, a bass speaker is almost omnidirectional so by placing it very close to a wall, all the reflected sound from it will be virtually in-phase with the sound from the front of the cabinet, resulting in a near doubling of level. Standing sub-bass cabinets in corners can produce a further doubling of level although corner mounting may cause boominess that would require EQ to compensate.

Stacking multiple bass cabinets also improves their low-frequency performance, as the array size become larger compared to the wavelength of sound being reproduced. For this reason placing a pair of subwoofers adjacent to each other may well be more effective than placing one on each side of the stage. That would also help avoid frequency dips at certain locations in the room caused by the possibility of phase cancellation when the bass energy originates from two separate sources. As these low frequencies are so hard to locate audibly, the subs can be stacked to one side of the stage if required without compromising the overall stereo image. Personally I find this better than putting the subs directly under the stage as powerful low frequencies can be distracting to the performers standing directly above them.

Compact Line Arrays

Line arrays have become a common sight at major auditorium and festival events these days, with a number of identical stacked speaker elements suspended vertically high above the audience, often with the lower cabinets pointed lower down to cover those seated near the front. The interaction between the cabinets affects the shape of the wavefront being generated, so instead of the spherical wavefront you'd expect to see from a lone single-driver cabinet, a vertical line array's wavefront is more like a cylinder, with wide horizontal dispersion and tightly controlled, limited vertical dispersion. To work exactly in accordance with theory, the drivers must also be spaced closer together than the wavelength of the highest frequency they are intended to reproduce, although this 'rule' seems to get 'bent' on a regular basis without compromising the result too greatly.

A vertical line array is characterised by a narrow vertical polar pattern that minimises sound wastage above the heads of the audience, but at the same time the horizontal dispersion is much wider than from a simple box. A major benefit of the cylindrical polar pattern is that the sound 'throws' further than from a single driver box. That's because a normal spherical wavefront has a sound intensity that quarters or falls by 6dB for every doubling of distance from that source. By contrast, the cylindrical wavefront of the line array yields an SPL that halves or falls by just 3dB for every doubling of the distance.

Of course, a line array would need to be infinitely long to perform exactly as described, but a practical line array still has a significantly wider horizontal dispersion and narrower vertical dispersion than a single-driver box. However the cylindrical dispersion pattern breaks down at low frequencies where the bass energy tends to become more omnidirectional, so small-scale line arrays tend to be reinforced with separate conventional subwoofers. The cylindrical dispersion pattern also breaks down when the listener is far enough away from the speaker array to perceive it as a 'point source', but that isn't an issue in most venues.

Comparison of spherical and cylindrical wavefronts.

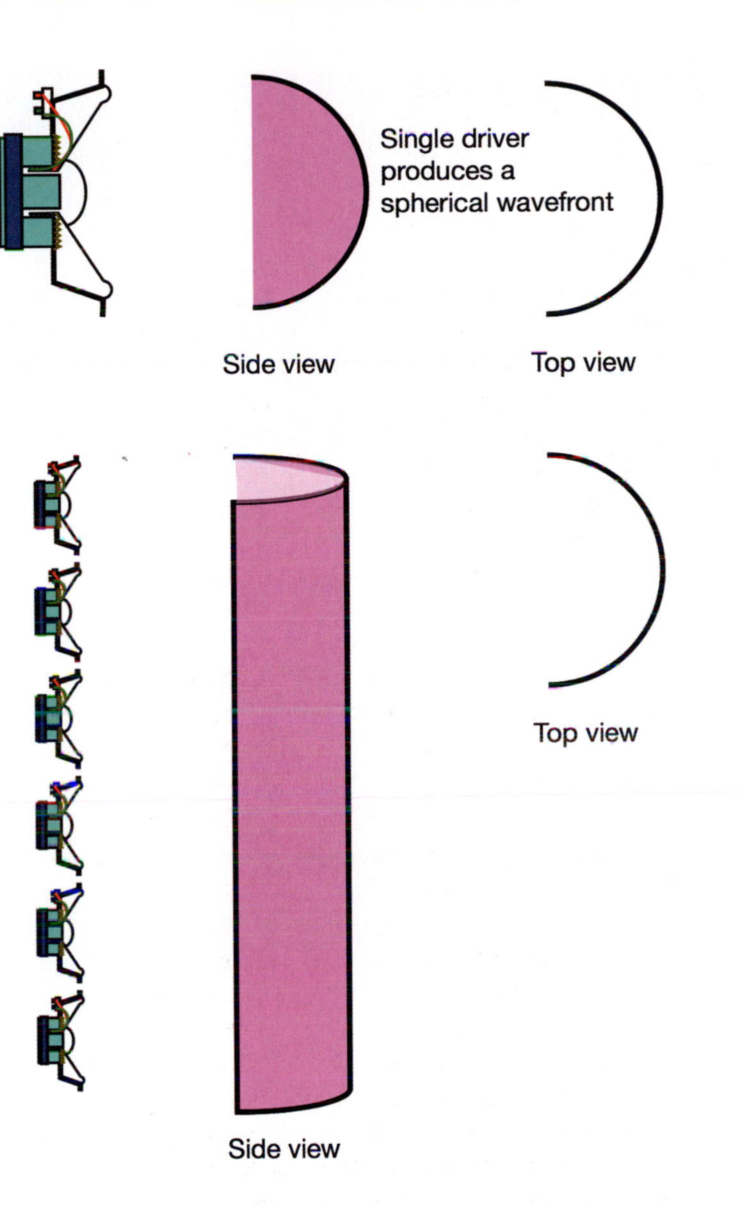

While the design and rigging of large scale line arrays can be very complex, a number of portable 'mini line arrays' have been produced that are particularly well-suited to the gigging musician playing small venues. All the ones I've seen combine a narrow column of small-diameter drive units with a separate subwoofer, and at least one of them bends the rules slightly by using a single tweeter mounted in a shaped horn or waveguide to try to match its dispersion to that of the multiple small drivers.

The vertical line arrays used for large-scale concerts are designed to exhibit wide horizontal dispersion, with precisely controlled vertical dispersion.

The line-array principle lends itself equally well to small-scale systems.

TIP: Small line arrays give good coverage and intelligibility in smaller venues with less of a 'dead spot' when standing between the speakers close to the front, and less of a volume difference between the front and rear of the audience area.

My experience with such small line arrays has been that they give good coverage and intelligibility in smaller venues with minimal visual disruption. Practical benefits that you will notice include less of a 'dead spot' when standing between the speakers close to the front and less of a volume difference between the front and rear of the audience area. Room colouration due to ceiling reflections should also be reduced, and, because of the wide dispersion angle, you may achieve a little more level before feedback becomes a problem. There's also the factor that smaller drive units can reproduce the critical mid-range more cleanly than a 12- or 15-inch two-way system, where the woofer and tweeter may struggle to 'meet in the middle', so vocal clarity, too, can seem better than with a conventional box. My own view is that small line arrays combined with one or two subs are the way to go in achieving good sound quality in smaller or difficult venues.

Passive Or Active?

Active speakers, with built-in power amps, have a clear advantage in terms of convenience, in so much as there's no need for external amplifier and crossover racks, and they can also be connected using standard XLR microphone cables rather than heavy-duty speaker cables. Passive crossovers of the type built into many passive two-way speakers are also electrically 'lossy', as explained earlier, so you need more amplifier power to get the same level from a passive system than from one that incorporates an active crossover. Another important consideration is that active speakers usually incorporate various forms of protection to limit the level being fed to the speakers and to monitor the system for overheating and other problems. External processors can offer similar speaker protection but require a degree of expertise to set them up correctly. A fully passive three-way system would require three separate stereo amplifiers, a stereo system processor (for the crossovers and protection) and six separate speaker cables, not to mention six XLR cables linking the processor to the amplifiers. In a typical pub gig that's a complication you can live without!

Smaller active systems, with all the power amplifiers located within a single subwoofer, are particularly vulnerable to amp failure because if the subwoofer's amp pack fails you lose everything. For a good compromise between flexibility, convenience and resistance to catastrophic failure, I'd therefore suggest either using a system that has two powered subs (in the case of the subs handling all the amplifier duties) or using a powered sub in conjunction with active tops. If one of the tops fails you can still carry on using just one of them (or press a monitor into service to fill in for the 'dead' side), and if the sub fails you can carry on simply with less low end.

Real-world Considerations

PA systems frequently have to work in less than optimal acoustic environments, and should always have the ability to cope with the occasional boomy or overly-reverberant room. Achieving acceptable results in the face of acoustic adversity requires a basic understanding of the relationship between the room acoustics and acoustical properties of the loudspeakers. Although it may be simply impossible to achieve ideal sound quality in some rooms, you can often improve on the situation by positioning the speakers where they will work best and by using an equaliser, such as a parametric or graphic EQ, to dip out those frequencies that coincide with the more offensive room resonances.

Critical Distance

In any room, the direct sound from a loudspeaker will become quieter as you move further back from it. This is as a result of the 'inverse square law'—a mathematical expression describing how sound waves get weaker as they move outwards from a point source, simply because their energy is dissipated over a wider area. Speaker systems that do not exhibit a point-source dispersion pattern, such as line arrays, don't follow the inverse square law, although their SPL will still get lower the further away you are from the speaker.

In contrast, reverberation, caused by sound bouncing around the room from surface to surface, does not change in level significantly with distance. This leaves us in the situation where the reverberant sound intensity is more or less constant throughout the room whereas the direct sound from the speakers will get quieter the further away you are. Eventually we reach a point when both the direct and reverberant sound are equal in intensity and this is known as the 'critical distance': beyond this point, the reverberant sound predominates and intelligibility suffers accordingly.

While the critical distance depends somewhat on the room acoustics, the amount of reverberant sound also depends on how much of the sound from your speakers reaches the walls and ceilings in the first place. It follows that if your speakers can be arranged to project more sound onto the audience and less onto the walls and ceiling, there will be a lesser proportion of reverberant sound and so the critical distance will be greater. The practical outcome is that you'll achieve greater clarity over more of the room. Both the choice and positioning of loudspeakers affect the outcome, as different speakers have different directional properties: for example, line arrays with their limited dispersion in the vertical plane will result in less sound bouncing off the ceiling.

TIP: As you move away from the sound system, you will eventually reach a point where both the direct and reverberant sound are equal in intensity—this is known as the 'critical distance', as beyond this point, the reverberant sound will dominate, greatly reducing intelligibility.

Plastic Fantastic?

A number of compact PA systems are now built in moulded plastic cases rather than wooden boxes. With the correct internal bracing, plastic boxes can be made relatively resonance free while offering certain practical advantages, such as moulded handles, port tubes, pole sockets and integral horn flares that can't work loose or rattle. The cabinets can also be shaped to help reduce internal reflections, although the non-rectangular shape can make them more difficult to stack along with other equipment during transport. It's also hard to seriously damage a well-designed plastic speaker cabinet as they're usually made from a very tough, resilient form of polyurethane.

The very best plastic-box designs can perform extremely well, whilst less well-designed or less well-manufactured ones can sound somewhat boxy and unfocussed. More problematic in my view is that significantly more sound seems to leak out through the cabinet walls than with a good wooden box design, and the muddy, boxy sound of the leakage can be very distracting when heard from the performing position. Some plastic speakers I've tested sound great out front, but the sound on stage is truly horrible, so you have to be very careful where you position them.

I have to admit to having mixed feelings about plastic boxes; they are tough, convenient and usually work well for compact systems with woofers up to 10 inches in diameter, but once you get to 12-inch drivers and above, you have to be very careful which model you choose, and experience suggests that, all else being equal, birch-ply wooden boxes still produce a cleaner sound. MDF boxes can also work well, although they tend to be noticeably heavier than birch-ply boxes. The same goes for plastic subwoofers—it is very difficult to design one that doesn't resonate.

The Class System

One of the biggest developments in live sound over recent years is the introduction of so-called Class-D amplifiers and 'switch-mode' power supplies. You don't need to be overly concerned with the technology behind these developments but you should try to understand their implications.

Switch-mode power supplies can usually be designed to automatically accommodate any international mains voltage, which is both convenient and helps avoid expensive accidents! They work by converting the incoming mains power to a much higher frequency, which allows very small and consequently light weight transformers to be used. Designed correctly they are very reliable, but if they do go wrong it is usually a case of replacing the whole power-supply board—the circuitry isn't something you can fix with some new rectifier diodes and a spare fuse, as was often the case with more conventional linear power supply designs.

Class-D amplifiers work using a system of 'pulse-width modulation' and are sometimes referred to [incorrectly] as digital amplifiers. As with switch-mode power supplies, these employ

Class-D power amps, such as this model from Peavey, are capable of delivering huge power outputs, combined with very low weight and very high efficiency.

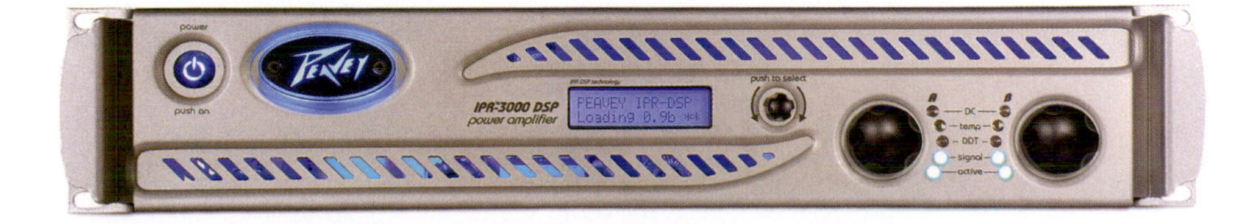

serious amounts of internal cleverness that makes them capable of delivering huge power outputs, combined with very low weight and very high efficiency. But if one goes wrong it's usually board swap time again, as the things are notoriously difficult to fix. Nevertheless, the high efficiency (which means much less unwanted heat) and light weight means that Class-D amplifiers are rapidly becoming the norm in high-powered PA applications and are also finding their way into many smaller systems. Now you can easily carry a 5,000-watt amplifier in one hand—not something you'd want to try with a traditional Class-AB design.

Chapter Two
Practical PA Systems

Choosing the right PA system to match both your needs and your budget can be a challenge, so the purpose of this chapter, having outlined the underlying technology in the previous chapter, is to take a closer look at the various types of system available. Whilst a major touring band invariably relies on a hired sound system with separate mix engineers for the front-of-house (FOH) sound and for the stage monitor mix, bands and artists working in smaller venues often provide their own PA system and either have their own mix engineer or indeed mix the sound themselves from on stage. Their requirements in a PA system extend beyond just 'will it do the job', to include affordability, portability and even expandability—can you easily add extra speakers, a subwoofer, or monitors at some time in the future? Fortunately, in today's market, there's now plenty of choice, but that wasn't always the case . . .

When I look back on the monstrous sound systems I used to set up in pubs only a few decades back I can't help but smile at how ridiculous it all was. We had folded-horn bass bins big enough to sleep in, even though the bass guitar and kick drum were already too loud even before being fed into the PA. Then we had four 12-inch midrange boxes with flared fronts and separate horn-loaded compression drivers perched on top. All this was powered by home-made power amps and crossovers built into ex-military grenade-storage boxes, and it all weighed a ton. When rigged, the gap between our two chipboard 'towers of excess' was just about wide enough for the audience (when there was one) to see the vocalist!

How things have changed! A modern, stereo PA system capable of delivering between a few hundred watts and a couple of kilowatts of power can now be small enough to fit into the back of a hatchback car or small van, and will range in price from the hundreds up to several thousands of pounds or dollars depending on the build quality and specifications.

The biggest part of the 'will it do the job?' decision, accepting that decent sound quality is a prerequisite, comes down to how loud you need the system to go and how much deep bass you need to put out. For example, if you're a solo artist or a duo working with just voice and guitar, and playing bars and coffee shops, there's no need for massive subwoofers capable of producing 120dB SPL at 30Hz. However, if you want to amplify an entire rock band or dance music, you'll definitely need to have a system that can reproduce low frequencies at high SPLs, and that means having at least one subwoofer.

Practical System Configurations

Bands playing smaller venues can also usually get away with a sound system with a less extended bass response, as the PA's main role will be amplifying vocals and perhaps reinforcing instruments such as guitar and keyboards. The bass player's own amplification and the natural acoustic sound of the drums is often loud enough in small venues, so, other than perhaps a bit of reinforcement for the kick drum, there may be no need to put these through the PA at all.

But what happens when you get offered a bigger gig at a larger venue? Unless you are in a position to borrow or hire compatible extra gear for the occasion, you may want to consider owning a more scalable system—one that allows you to take fewer speakers for smaller gigs and more for the bigger ones. Many small two-way speakers are designed to facilitate use either as main speakers or as monitors, which helps when you come to plan a flexible system, as you may be able to use them as your main PA in smaller venues but then press them into service as monitors when you play a venue that needs your larger speakers. Furthermore, some two-way speakers that might normally only be suitable for smaller venues when used

Modern PA designs offer high output and controlled dispersion from systems that you can transport in the back of a car!

ACOUSTIC AMPS AS PA

Some solo artists and duos may be able to manage without having a traditional PA at all, as most acoustic-guitar combo amplifiers have a separate microphone input allowing them to function as a complete 'mini PA' amplification system in their own right. The majority also have a line-level DI output that allows a second powered speaker to be connected, so adding something like a 10-inch, two-way PA speaker can often provide perfectly adequate coverage even for slightly larger pub and coffee-shop gigs. Popular names in acoustic guitar combos include AER, Fishman and Schertler, although many electric-guitar amp manufacturers, such as Vox, Marshall, Yamaha and Roland now also offer acoustic models.

on their own, can be teamed with a subwoofer to extend their capabilities sufficiently to allow them to take on a bigger job.

Most basic, 'easy-to-use' PA systems are designed with portability in mind—the Yamaha Stagepas range being a good example of what can be achieved. Plastic-cabinet, passive two-way speakers are fed from a small powered mixer (the power amplifiers are built into the mixer) that is designed to pack away

Self-contained amps designed for vocal and acoustic guitar use have the same flat-response, full-range specification as PA systems.

into a recess in the back of one of the speaker cabinets for transport. A matching compartment on the other speaker may be used to store microphones and cables, so you can walk into a venue with a cabinet in each hand, carrying your entire PA . . . other than mic stands . . . and speaker stands . . . and mains extension leads! Fender's Passport system is another ingenious portable PA solution, with two speakers that clip together to form one suitcase-sized unit.

Despite offering a few hundred watts of power, and being capable of SPLs in excess of 125dB, these systems clearly aren't designed to handle significant levels of bass, but they are very well suited to solo acts and duos where there's only a need to amplify guitars, maybe a keyboard, and voices. The more powerful variants are also capable of delivering adequately loud vocals in a 'band with drums' context at smaller venues. Some even have connectors built-in for adding an optional sub-woofer, which would allow some back-line amplification, or a kick drum to be amplified. The main limitation of such systems

▲
Highly portable, integrated PA systems such as this Yamaha Stagepas system are designed to be particularly easy to use.

is that they are seldom designed to be easily expanded: for example, whilst their mixers may have additional line outputs that can be used to drive more powered speakers, many have no provision for driving stage monitors, and not all of them have the ability to properly integrate a subwoofer.

Satellite Systems

We normally think of separate subwoofers being used to allow larger PA systems to handle full-range audio material at high levels, but, at the small-scale end of the market, there are compact systems that use what is, strictly speaking, a separate 'woofer' rather than a true subwoofer. In such systems the purpose of the woofer is to extend the range of a pair of fairly small satellite speakers, usually fitted with six-or eight-inch main drivers, to create an overall system response that can equal what you might expect from a pair of 12- or 15-inch two-way speakers. The real low end of these systems won't go down as far as a true sub, probably cutting off in the 45 to 60Hz region, depending on size.

In these systems, the bass speaker, although often described as a 'sub', is not an optional add-on but a key component. The satellite speakers are not designed to work without it and can therefore be made smaller, as they don't have to handle frequencies below 120Hz or thereabouts. This arrangement is essentially a three-way system, in which the top cabs handle the mids and highs in stereo while the bass speaker handles all the lows in mono. Their compact physical format is an advantage in very small venues as the compact satellite speakers won't obstruct the audience sight-lines and there's plenty of flexibility in positioning the sub. Although you may be able to connect a microphone directly to this kind of system in an emergency, you would generally use a separate mixer.

HK Audio's smaller LUCAS systems fall precisely into the 'satellite plus bass speaker' category whereas their larger models have true subwoofers that extend further into the deep bass region. A number of manufacturers now build systems following this format, with the smaller ones being fine for solo artists, duos, acoustic-instrument based acts and also for vocal-only PAs for small-venue band use. The larger examples, however, are essentially full-range PA systems that will happily accommodate all the back line, including bass guitar and kick drum. One thing to bear in mind, however, with satellite-and-sub systems that use passive top cabs with all the power amps built into the sub, is that if the amplifier pack fails, you would lose audio from the whole system.

It is difficult to define exactly where the woofer-plus-satellite systems stop and true full-range systems take over, as the 'two-tops-plus-bass-speaker' format comes in a whole range of sizes, from shoebox-sized satellites to a pair of 12-inch tops augmented by a 15- or 18-inch sub. As a general rule, I'd say that if the tops are designed to be usable without the added bass speaker, then you have a true modular PA system, whereas if the manufacturer suggests that the tops should never be operated without the subs, then you have a satellite system. In my experience, full-range tops with 10-inch drivers and above can usually turn in a good performance without relying on a sub when used as a vocal PA.

A typical compact satellite-and-sub system.

Modular Systems

I've actually handled quite sizeable outdoor musical events using just a pair of 12-inch tops and a single 15-inch subwoofer, so the 'tops-plus-sub' format can be very capable, and flexible—you just pick the size and specification you need. Adding a second sub can also extend the usefulness of the larger systems for when you need to reproduce higher levels of bass. It is also possible to double up on the number of top cabs, so this approach really is ideal if you need a system that can be tailored to different venues.

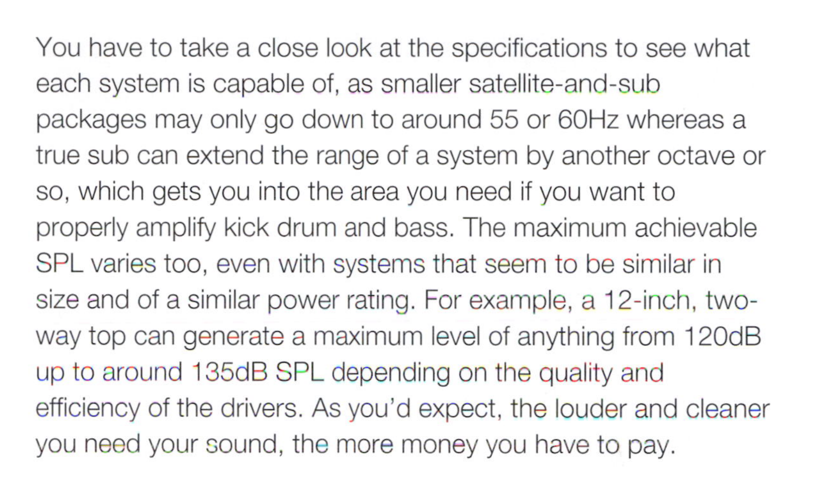

You have to take a close look at the specifications to see what each system is capable of, as smaller satellite-and-sub packages may only go down to around 55 or 60Hz whereas a true sub can extend the range of a system by another octave or so, which gets you into the area you need if you want to properly amplify kick drum and bass. The maximum achievable SPL varies too, even with systems that seem to be similar in size and of a similar power rating. For example, a 12-inch, two-way top can generate a maximum level of anything from 120dB up to around 135dB SPL depending on the quality and efficiency of the drivers. As you'd expect, the louder and cleaner you need your sound, the more money you have to pay.

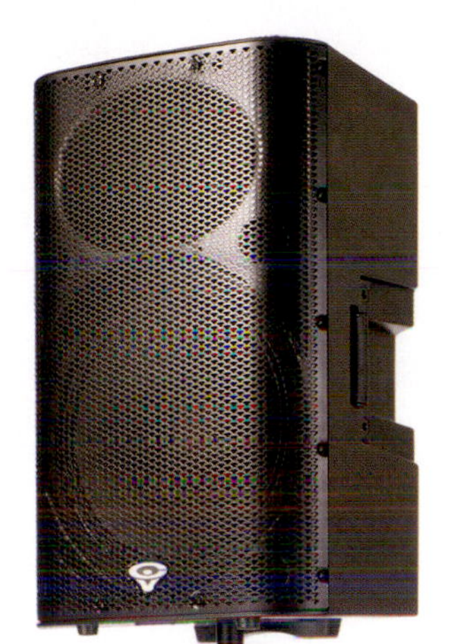

Larger top cabs can often be used without the sub-woofer, creating an easily scaleable, modular PA system.

The (Only Slightly) Techy Stuff

Where there isn't a need to reproduce very high levels of bass, two-way systems are very common, employing an eight-,10- or 12-inch (and sometimes as large as 15-inch) speaker to cover the lows and midrange, combined with a horn-loaded tweeter to cover the highs. The crossover point is usually set between 2.5 and 3kHz, depending on the drivers being used, and it is common for the cabinet to be ported to increase the low-frequency efficiency. Such systems may be active, with power amplifiers and crossovers built into the speaker cabinet, or passive, with a passive crossover built into the speaker cabinet to split the incoming full-range signal from an external amplifier to feed appropriate signals to the LF and HF drivers.

Common cabinet materials are wood, usually plywood or MDF, or plastic—in plastic cabinets, the tweeter's horn flare is often moulded as an integral part of the cabinet, along with the handles, feet and other fittings. Many cabinets also have a trapezoidal cross-section or cut away rear corner allowing them to be positioned at an appropriate angle to be used as wedge monitors.

Modular two-way main speakers can easily be teamed with one or more subwoofers: the simplest setup being to use a single active subwoofer that takes the stereo feed from the mixing

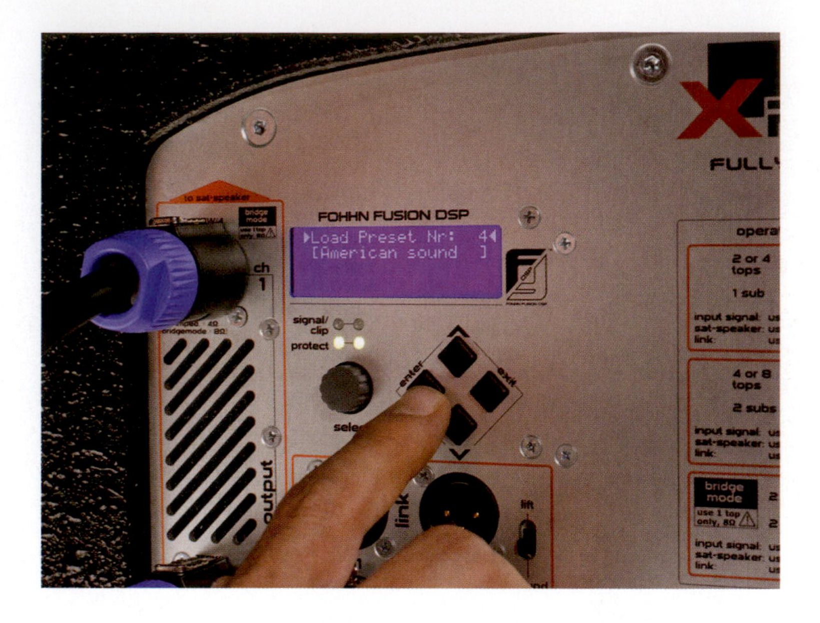

Sophisticated active subs, such as these built by German company Fohhn, may include the option to choose from a range of stored settings to match different configurations of main speaker.

desk and then outputs a high-pass-filtered (around 120Hz) line-level stereo feed to drive the main speakers. More sophisticated active subs may include the option to choose from a range of stored settings to match different configurations of main speaker. These presets may combine crossover settings with some overall EQ and appropriate limiting.

Because the sub is taking care of all the lows and the sub's crossover generally includes high-pass filtering for the feed to the top cabs, the main speakers can work more effectively at mid and high frequencies, as they no longer have to deal with the power-hungry bass end of the spectrum. The long wavelengths involved in bass reproduction make the source of deep bass almost impossible to locate in an enclosed space—human hearing systems use almost exclusively mids and highs to work out where a sound is coming from—so it is perfectly permissible to position the bass speakers away from the mid/high boxes, if space dictates.

One or two 15-inch subs will add a lot of low-end punch, although for smaller venues a 12-inch or even a 2 × 10-inch sub may be adequate. Whilst 18-inch subs do shift more air, larger diameter speakers tend to require larger enclosures volumes, so their physical size tends to make them less attractive in situations where band members have to transport and rig their own PA!

Line arrays using multiple small drivers, paired with sub-woofers have become a popular solution for portable systems.

Mini Line Arrays

'Mini' line arrays are becoming increasingly fashionable once again, typically employing multiple small cone drivers (typically three or four inches in diameter) mounted in a narrow, lightweight enclosure. Small drivers allow the columns to be made very compact, especially as they have no requirement to handle bass frequencies. All the systems of this type that I've tried must be used with separate subs and in most cases the necessary power amplifiers are built into the subs, though HK Audio's Elements system also offers separate power amps that slot onto the bottom of the speakers. The Elements system also allows multiple modular line-array speaker boxes to be clipped together to form a longer line.

I use a small Fohhn line array system on a regular basis for smaller venues, and I can carry the 12-inch powered sub in one hand and both line-array columns in the other—definitely an improvement over my chipboard monoliths of old! For bigger gigs the same tops will work with a 15- or 18-inch sub. Designers are always looking for ways to make setting up simpler and several of the mini-line arrays I've looked at have all the necessary connections between the subs and tops built into the cabinets and mounting poles so that the only physical

The L1 system from Bose is unique in having been conceived as being able to operate as PA, monitors and backline, all at the same time.

connections that need to be made are two audio cables for the input and one power cable to the subwoofer rear panel.

Not only are these little systems easily portable and visually unobtrusive, they also direct more of the sound where you need it to go: their disadvantage is that with fixed-sized line arrays there's no simple way to expand them, as using two line-array cabinets side by side seriously compromises their directional properties. Even with modular systems such as the HK Elements range, there's a limit to the maximum system size. Big touring-PA companies build their large line arrays from multiple identical boxes, so they have a lot of flexibility when it comes to system size, but compact line arrays are far more restricted in their ability to be expanded.

Bose offer their L1 narrow column-speaker series, which looks like a mini line array augmented by a small subwoofer. However, their approach differs from the usual line array in that their drivers are angled to the left and right of centre to produce a very wide dispersion pattern. This allows them to be used in close proximity to the performers without aggravating feedback issues, and in many cases they can even be used behind the performers to provide both monitoring and front-of-house sound for smaller venues. They seem to be particularly popular with ensembles that don't require high SPLs but who do require natural-sounding level augmentation.

Active And Passive Issues

Depending on the size and design of the PA system, the crossovers between the mids and highs may be either active or passive, although the crossover between subs and top cabs is almost always active. There are systems that place all the power amps in the sub, but using either passive tops with an external amplifier or active tops with their own internal amplifiers provides a much better safeguard against catastrophic system failure. My own view is that, if you are transporting your PA yourself, active speakers are far more attractive than passive speakers, saving you the necessity of carrying and rigging external power amplifiers and crossovers and a lot more cabling.

It is prudent, however, to have an emergency strategy based on what you might need to get you out of trouble if a key piece of PA equipment should fail on the night. This might include having active monitors that can be redeployed as main speakers, speakers capable of accepting mic level signals (in case your mixer fails) or simply carrying an extra speaker or two. I managed to salvage one small gig where the whole of one side of the PA failed by using a small Mackie personal monitor with a single four-inch speaker to fill in for the missing side!

Smart Speakers

It has been common for some years now for manufacturers to equip their powered loudspeakers with one or more microphone inputs to allow them to be used in situations where a full-blown mixer is not needed, or as an emergency fix if your mixer fails. However, on-board processing and facilities have become a whole lot more sophisticated over the past few years with a new generation of active loudspeakers that use DSP technology, not only to handle crossover duties but also to provide additional features such as speaker voicing, safety limiting, anti-feedback filtering and even reverb and effects to be added. Mackie's DLM series, for example, offers vocal effects, anti-feedback, EQ and preset voicings for a number of different applications such as music, spoken word, general PA, and so on.

Line 6, famous for their pioneering work in the field of guitar-amp digital-modelling products, are also pioneering new ideas in the PA world. Their StageSource speakers can be connected as part of a digital network so that they are 'aware' of each other and able to adapt their parameters according to how many of them are in use and whether or not there is a sub in the system. They also sense whether they are upright in 'PA mode' or on their sides as monitors, and change some of their parameters accordingly. They can also be voiced for PA use or as guitar back-line speakers—their on-board effects include acoustic modelling taken from their Variax technology, helping a DI'd acoustic guitar to sound more like a studio miked guitar. When two speakers are networked, their mixers can be combined, and the whole speaker system integrates fully with the company's StageScape digital mixer.

A few years ago I wrote a leader column in *Sound On Sound* magazine suggesting that what was needed for bands playing smaller gigs was an active speaker system that also included some form of monitoring for the performers plus basic LED lighting. This was based on my personal experience of playing pub gigs where there's often no space to set up lighting stands and also very limited space for monitors. Putting all these essential components in one box means you have no more wiring than would normally be required for a pair of active speakers and it would be a whole lot easier to transport and rig.

Well, the idea can't have been too crazy as Studiomaster have now produced a product exactly like that in the form of their Starlight speaker system! It will be interesting to see who follows suit and whether the idea catches on.

StageSource speakers from digital modelling pioneers Line 6 bring the digital networking of high-end concert systems down to the self-operator level for the first time.

Chapter Three
Live-sound Monitoring Options

While many all-acoustic musical ensembles are easily able to achieve a natural balance, with each performer able to hear both themselves and everything that their colleagues are playing, the landscape changes as soon as amplification enters the picture. Amplification has made it possible for instruments to be combined in a musically meaningful way when, in the acoustic world, their relative volumes would have made coexistence impossible—for example, a nylon-string classical guitar alongside a drum kit. In fact, in our amplified world, just about anything can now be combined, but this presents us with a new set of problems because, while the out-front sound might be perfectly balanced, the musicians on stage often need help just to hear themselves and each other.

PA systems, are designed to be fairly directional, particularly at higher frequencies, so unless any artist-monitor speakers are provided, what's heard on the stage tends to be a rather indistinct sound dominated by low frequencies, which can actually be more distracting than getting nothing back at all. Throughout the early years of amplified musical performances, right up to the early 1970s, on stage monitoring remained a rarity for all but top-line professional performers. Fortunately, these days, stage monitor systems come in many different guises and in formats to suit all types of venue.

Floor Wedges

One of the earliest solutions, and one that's still widely used today, was to use a number of individual speakers across the front of the stage pointing back at the performers. These are usually wedge-shaped so as to direct the sound upwards at a fairly steep angle, and fed from pre-fade aux sends on the mixing console. This arrangement enables the engineer to set up as many individual monitor mixes as the mixer has pre-fade sends, with each one feeding its own monitor speaker or chain of speakers. While the pros normally use a dedicated monitor mixer located at the side of the stage, in smaller venues it is usual to rely on the front-of-house mixer's pre-fade sends, although few compact mixers have more than four.

Like normal PA cabinets, monitor speakers may either be active, meaning they have their own power amplifiers built in, or passive, fed from separate power amplifiers. While it is possible to get by in smaller venues with everyone hearing the same monitor mix, or even a duplicate of the front-of-house sound, singers especially often need to hear more of themselves than everything else, while drummers and bass players are often more concerned with hearing each other.

In the early days of stage monitoring, when the technical issues were perhaps not so well understood as they are now, there was always a temptation, outside professional circles at least, to put the cheapest and weakest-sounding speakers into floor wedges because, after all, the audience wasn't going to hear them! However, this can cause more problems than having no monitors at all as a poor-quality monitor speaker with an uneven frequency response and uncontrolled directivity makes keeping feedback under control especially difficult.

Mini Personal Monitors

Another approach that works very well for smaller venues is to use small 'personal monitors'. These are small speakers—usually self-powered and often little larger than a shoe box—that can be mounted on a mic stand and placed close to the performer. Some units can deliver as much as 150 watts

Wedge monitors across the front of the stage, angled steeply upwards so they are pointing at the ears of the frontline performers, rather than their knees!

through a four-inch speaker, aided by a 150 to 200Hz low-cut-off point. On their own, these tiny monitors do sound a little thin at the low end, as you'd expect with such a high cut-off frequency, but the missing lows tend to be filled in by bass leakage from the front-of-house speakers in a gig situation, making them a very practical choice for smaller venues. Galaxy Audio were one of the first companies to pioneer this monitoring solution with their Hot Spot series, but others have followed.

TIP: Compact, personal mini-monitors tend to sound very weak at the low end when heard on their own, but during performance the missing lows are filled in by bass leakage from the front-of-house speakers, making them a very practical choice for smaller venues.

Recognising that not all small mixers provide enough pre-fade sends to set up individual monitor mixes, many of these personal monitors now include a mic preamp and mic output, often along with some on-board EQ and the ability to mix in a second input. This makes them very flexible and easy to use—if all you need is to hear some of your own vocals, you just plug your mic into the monitor and then send the output from that on to the main PA mixer.

'Personal monitors' such as these units from Mackie and TC Electronic, are designed to be mounted very close to the performer and also give them some basic monitor mixing options.

Side-fills

Cross-stage side-fills are essentially PA speakers placed at the side of the stage pointing across, and they often carry something close to the front of house sound in order to make the overall on-stage sound more 'normal' and comfortable for the performers. They were originally only to be found at larger events, used in conjunction with individual wedge monitors, but with the advent of compact powered monitors, side-fills can be a practical way of providing general monitoring at smaller gigs so that everyone in the band gets to hear enough of what's going on without needing individual monitor speakers. Unless the drummer needs dedicated monitoring at the back of the stage, a pair of side-fills plus a main vocal monitor may very often be all that's needed.

The large PA stack to the far right of the picture points across the front of the stage rather than out into the audience, and is being used to provide 'side-fill' monitoring to supplement the wedges.

In a small venue where space is tight, a simple pair of cross-stage monitors can be the best solution.

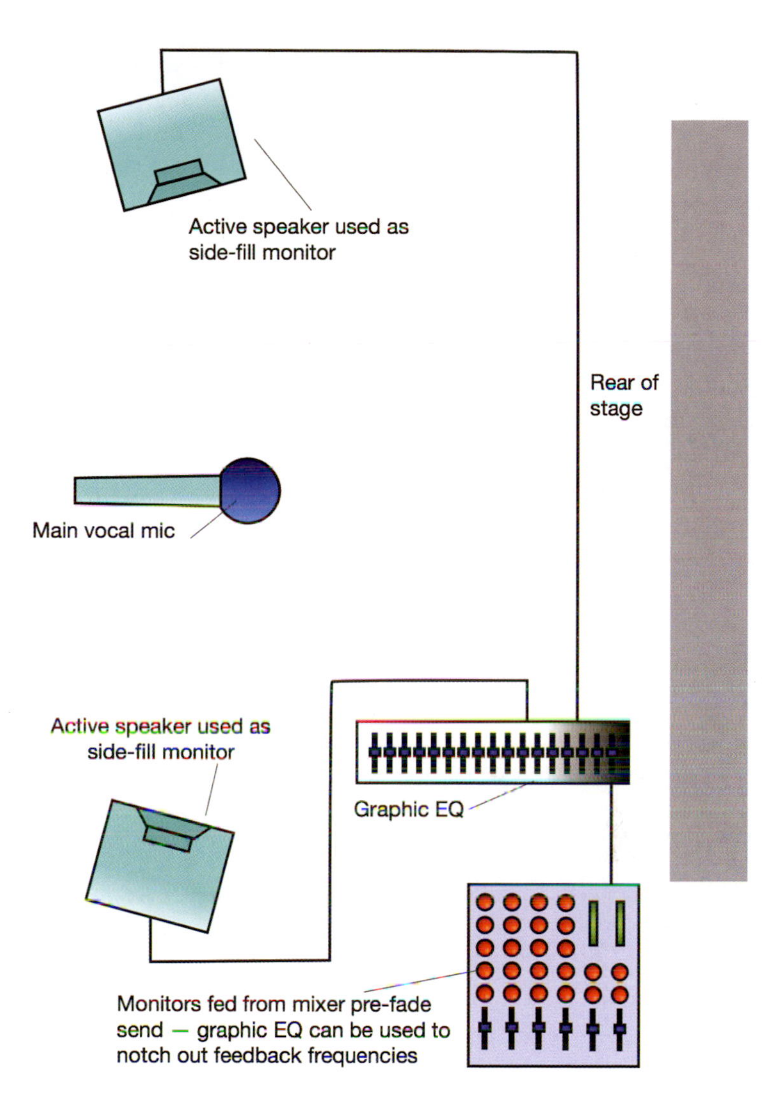

If you have enough aux sends to have a dedicated mix for the side-fills, it may necessary to put more of the vocals into balance, rather than simply duplicating the front of house mix, especially if you don't have provision for a separate vocal monitor, as most vocalists like to hear themselves as clearly as possible to ensure they are pitching accurately. Any monitor mixes sent to full-range cabs may benefit from having some of their low-end rolled off if possible to compensate for bass leakage from the front-of-house speakers, which has the added benefit of keeping the on-stage sound a little cleaner, too. I have found that using cross-fills in conjunction with separate vocal monitors works particularly well for outdoor events where there are no significant reflections to compromise the sound.

In-Ear Monitoring

Big-name touring bands have been using in-ear monitoring for many years now, but the technology is starting to trickle down to smaller acts playing more modest venues. The concept is simple—monitor mixes are fed to high-quality earpieces worn by the performers, and because there's no need to have conventional wedge monitors as well, there are no monitor-related feedback problems and there is no monitor spill into the on-stage mics, cleaning up the FOH sound to a considerable degree. Also, as properly fitted earpieces can block out much of the sound from the backline, it is often possible to perform with a significantly lower level in your ears than with wedges, which is good news when it comes to preserving your hearing.

How much sound is blocked out depends on the type of earpiece used, however. Off-the-shelf models usually come in the form of a plug with soft rubber flanges that seal against the sides of the ear canal as the earpieces are inserted, but, in practice, these generic plugs only provide a modest degree of sound isolation from the outside world. Custom-made models, designed to exactly fit the user's ears, often referred to as 'ear moulds', are far better: these are created by making a soft rubber 'master' mould of the wearer's ear canal, from which sets of earpieces can be manufactured as required. Many companies now offer a custom-mould service and the process is both quick and painless. The transducer inserts placed into the resulting custom earpiece range from basic, single-driver units similar to something that might be used for general music listening, through to complex multi-armature types capable of greater fidelity and much higher sound levels.

How you get signal to the earpieces depends on how much mobility you require on stage: the most flexible option is to use a wireless receiver pack, although this adds to the cost and complexity. Guitar players who use a conventional cable rather than a radio transmitter may find that a conventional wired

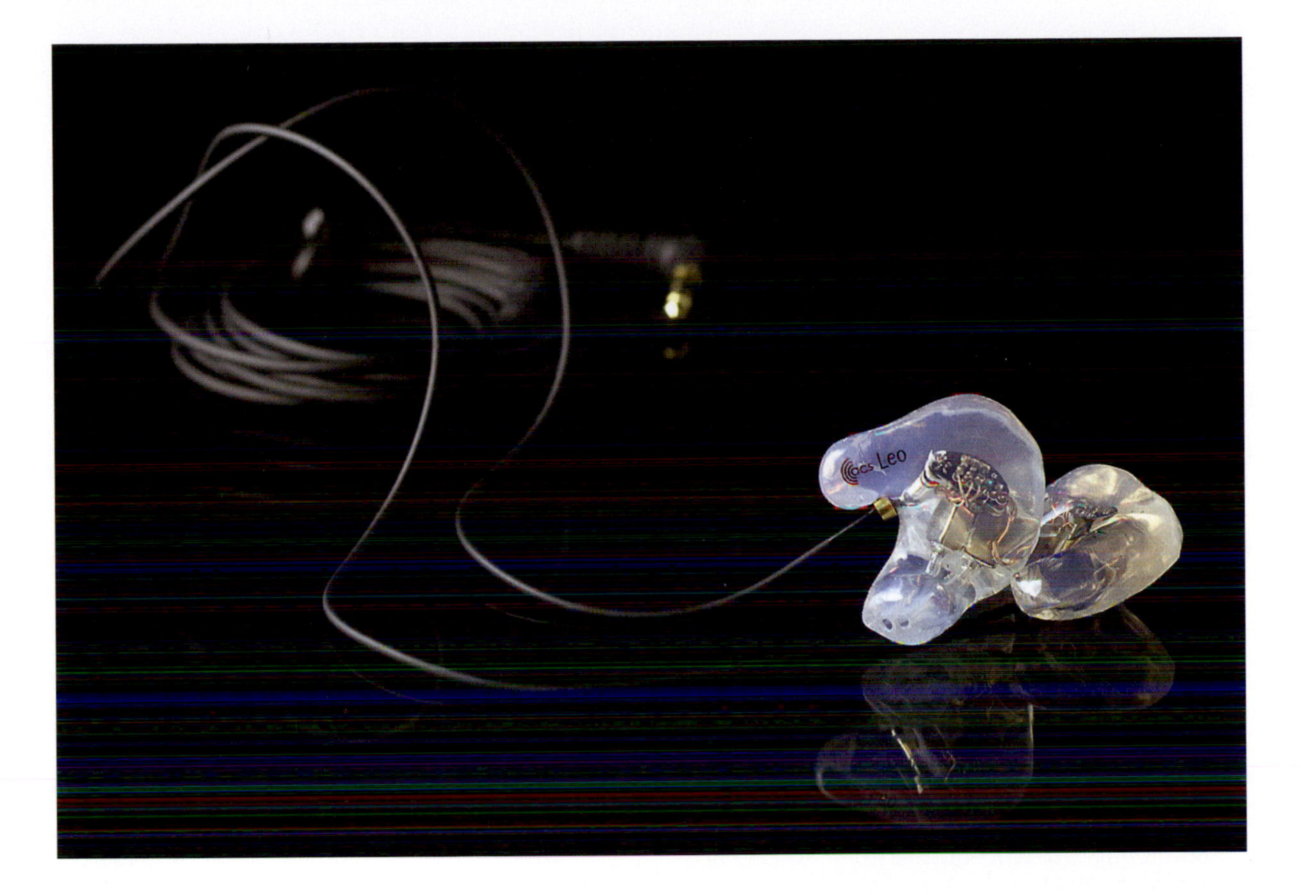

In-ear monitoring works best when custom-moulded earpieces that fit the wearer's ears perfectly are used, as this enables all external sound to be excluded, giving complete control over the level and balance heard by the performer.

connection works perfectly well for them—the cable could even be fixed to their guitar lead. In all cases, however, the headphone amplifier driving the in-ear monitors must be equipped with a safety limiter to prevent damaging levels of sound reaching the performer, either through 'volume creep' in performance or somebody accidentally turning the wrong knob.

User reaction to in-ear monitors varies somewhat, with many performers in large-scale shows describing them as a 'necessary evil'. With most of the external sound shut out, all you are hearing is whatever is in your monitor mix, so you can't just move closer to your amplifier when you can't hear your guitar well enough, as you might when using conventional monitors. Similarly, unless you have audience mics fed into your mix, you may feel isolated from any audience reaction. For singers, there's also the issue of hearing your own voice via both the sound from the in-ears and from bone conduction within your head, which can really change the perception of how you sound: some users find that it significantly affects their ability to pitch correctly. Most important of all, however, is that you must really trust your monitor engineer, or control your mix yourself, as an in-ear monitor wearer lives or dies by their monitor mix!

A recent development in in-ear monitors has been the inclusion of miniature microphones—on the in-ears or the wireless pack—whose signal can be mixed into the monitor feed by the performers themselves, providing some ambient sound to prevent the feeling of isolation.

Personal Monitor Control

The simplest form of personal monitor control is essentially a small analogue mixer that allows the performer to control both their own monitor level and the relative level of themselves as heard against the rest of the monitor mix. In many cases simply being able to balance your own voice or instrument against the main front-of-house mix is perfectly adequate.

Much greater levels of sophistication are available in the form of digital personal monitor mixing systems, usually made up of multiple individual mixing stations networked together with the FOH mixer via Ethernet. These allow each performer to set up their own monitor mix based on a number of sub mixes, or sometimes even all the available pre-fade sends from the desk—you can even add vocal effects such as reverb or delay to the monitor mix to help with pitching. In general, I think these are probably a bit over-the-top for bands playing small venues, not to mention quite costly, especially when simple, affordable analogue solutions are available.

There are several compact multi-purpose digital mixers that can be remotely controlled via wi-fi by personal computing devices, such as tablets or smart phones. This opens up the possibility of individual performers being able to tweak their monitor balance during a performance using their smart-phone or tablet mounted on a mic-stand bracket. Most of these mixers offer some form of access restriction so performers can be allowed

Networked digital live sound systems facilitate the inclusion of sophisticated personal monitor mixers, allowing performers to precisely tailor their own balance and level settings.

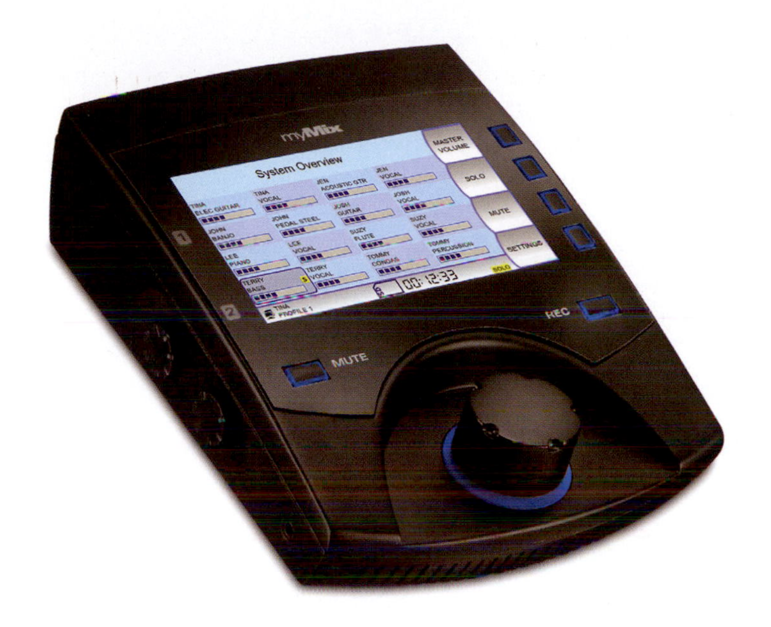

to change their own monitor mix but not affect the main mix or anyone else's monitor mix. This approach can be particularly useful for in-ear wearers who don't have a separate monitor engineer. However, where conventional wedges or compact monitors are being used, there's a very real risk that a performer may turn up their own mic's monitor level far enough to provoke feedback.

Monitors and Microphones

Stage monitors should, whenever possible, be placed in the 'dead spot' of a microphone's polar response. For a cardioid microphone, this means directly behind the mic, not just in the horizontal plane but also in the vertical, which may mean tilting the mic upwards at an angle to comply. In this respect hyper-cardioid mics may be easier to deal with as their zone of lowest sensitivity lies around 45 degrees off-axis to the rear, which means that the mic can be left in a more natural near-horizontal position. Check the spec sheets for your stage mics to find the exact angle and maybe even make up a cardboard or wooden triangle that you can hold alongside your mic to help you line it up precisely until you've got it fixed in your memory. Also be

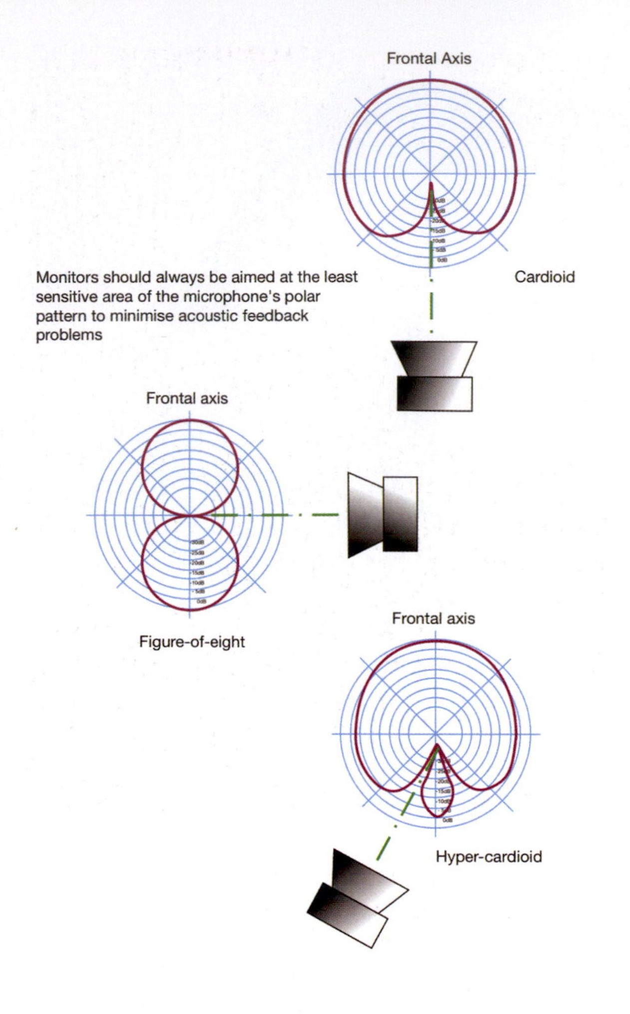

An awareness of the microphone's angle of maximum rejection helps to minimise feedback problems caused by nearby wedge monitors.

aware of any hard surfaces, such as walls or large pieces of equipment, that might reflect sound from the monitors back into your microphone as this too will reduce the level that can be achieved before feedback becomes a problem.

Reducing Monitor Feedback

It is quite common to use graphic equalisers in the monitor speaker feed to notch out any problem frequencies—these can be found during system set-up by increasing the monitor gain until feedback occurs and then using the appropriate graphic-equaliser band to tame the offending frequency, increasing the

Trimming problematic feedback frequencies without significantly compromising the overall tonality of the audio requires either a third-octave graphic EQ or the precision of a parametric EQ.

gain still further and notching out the next howlround frequency and so on (a process known as 'ringing-out'). You'll need a third-octave graphic equaliser to take out feedback frequencies without also affecting adjacent bands to an unacceptable degree. The integral octave-band graphic equalisers found on many powered mixers are close to useless for the purpose—it's a bit like being asked to carry out brain surgery with a soup ladle!

A simpler, and these days often more effective, approach is to use an automatic feedback eliminator, either just in the monitor feed or on both monitors and the main PA feed. These devices will be discussed in more depth in Chapter 8, but essentially they comprise a number of very narrow, digital filters combined with a spectrum analyser. Whenever the processor detects feedback it automatically deploys one of its filters to reduce the gain at that precise frequency. The filters are much narrower than the bands of a graphic equaliser, often just a fraction of a

semitone, so their effect on the overall tonality is much less severe than using a graphic EQ. While not a 'magic bullet' solution for feedback, these devices can claw back some useful headroom, as well as preventing that irritating 'on-the-edge-of-ringing' sound you hear when something is set close to feedback.

However, the feedback has to build up to an audible level before the system can recognise it and counteract it, so it is best practice to set up maybe half of the filters with a conventional 'ringing-out' process, leaving the rest as floating filters to deal with any 'random-event' feedback that might occur during performance—particularly important if the vocalist likes to wander around using a hand-held mic. Most feedback eliminators allow the user to decide how many filters will be 'locked' after ringing-out and how many will be left free to roam.

▲ The two Dbx feedback eliminators in the centre of this rack are working hard, judging by the number of filter lights illuminated.

Chapter Four
Audio Mixers

Mixers, however simple, form a vital part of any live-sound system. Not only does the mixer provide the means to balance and equalise a number of different sources, it also offers a means of routing signals to different destinations, such as the main PA system, monitor speakers, external effects and processors, or internal effects. Building a mental map of the signal flow within your mixer really lets you get the most out of it, and certainly helps you troubleshoot the system when you are wondering why you can't hear anything! Despite the general trend towards ever more affordable compact digital mixers for live sound, it is still helpful to fully understand the workings of an analogue mixing console, as the analogue operational paradigm is mirrored closely by most digital mixers.

Channel Building Blocks

A typical mixer is made up of multiple identical 'building blocks'—the majority of these will usually be input channels, each fed from an individual audio input. The channel includes the necessary microphone- or line-level preamplifier (sometimes both), an equaliser (EQ) section, and some means of level control—usually a linear fader. Further controls, called auxiliary sends allow some of the signal to be sent to monitors and effects. A simple mixer design would send all its channels directly to its main mix output via something called a 'mix bus' (essentially, a place where the channel outputs are mixed together), but more sophisticated mixers may include some more user-configurable routing involving additional buses, the reason for which will be explained later in this chapter.

The input channels on mixers of all sizes tend to be made up of the same elements—gain, EQ, auxiliaries, output routing, pan and fader.

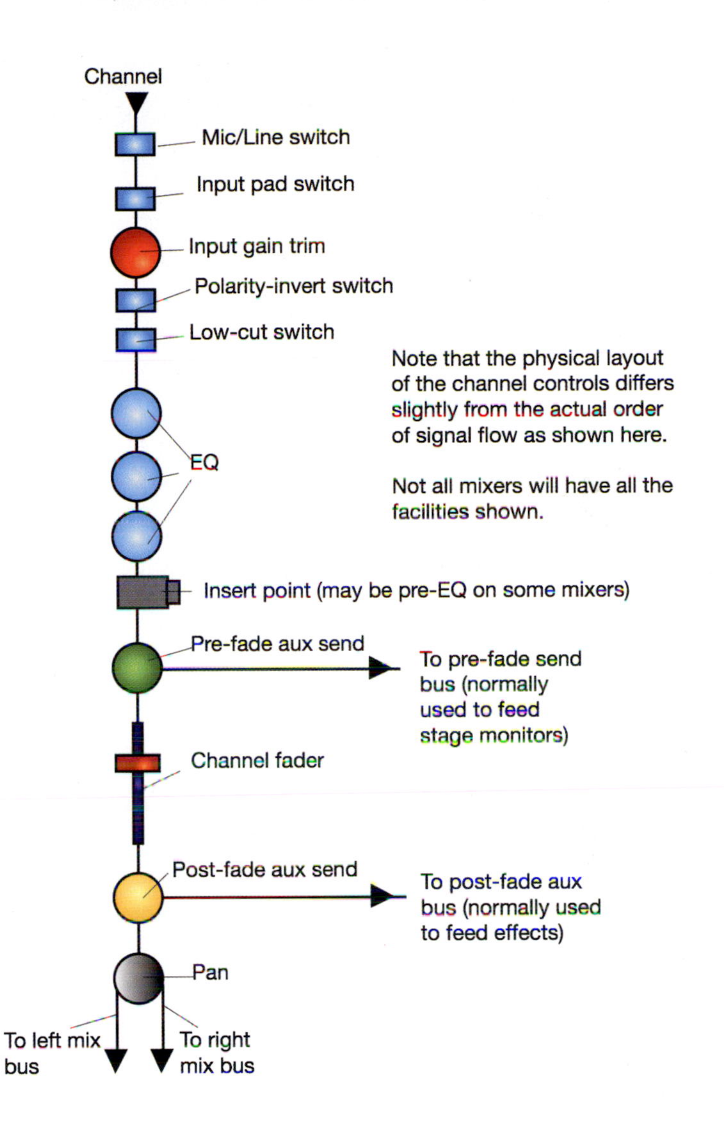

Mic And Line Inputs

Internally, analogue mixers can only work within a specific range of signal levels determined by the type of circuitry used. The goal of the mixer's input stage is to amplify the input signal to a level where it falls within the mixer circuitry's optimum operating range—well above the noise floor, but below the point where overloads and clipping occur.

Mixer inputs may be optimised for either microphone-level or line-level signals, with appropriate input impedances for these sources (see Chapter 13). 'Line level' is actually a bit of a vague term, but it can generally be taken to refer to signal levels of a volt or so, rather than the few millivolts that a microphone produces. Mic level signals need rather more preamp gain to bring them up to the mixer's internal operating level.

Bells And Whistles

That brings us into the murky but unavoidable world of the decibel, in all its forms—dB, dBu, dBm, dBv and dBV—as they all relate to standard audio operating levels. The decibel (dB), as the name suggests, is a tenth of a Bel: the term Bel taking its name from that famous pioneer of telephony, Alexander Graham Bell. On its own, the dB doesn't relate to a fixed signal level but rather is a means of expressing the ratio between two signal levels. The decibel is useful for audio because it operates on a logarithmic scale, like our ears, which allows it to express a very wide range of values in a meaningful way.

An analogue mixer's level meter is calibrated so that the nominal operating level is shown as 0dB or 0VU (Volume Units), leaving a reasonable amount of 'headroom' or safety margin above that level. If a signal is lower than 0VU, it is said to be minus so many dB, and a signal higher than 0VU may be expressed as being plus so many dB. Fortunately, you don't need to do many actual calculations involving dBs, but knowing some of the more common ratios can be useful. For example, a voltage amplifier having a gain of 60dB makes the input signal 1,000 times larger, while a dynamic range of 100dB means that the largest signal voltage a circuit can handle is 100,000 times bigger than the smallest signal it can handle. A voltage increase of 6dB means the voltage has been doubled, but the decibel can also be used in the context of audio power, where a doubling of power is +3dB and a halving of power is -3dB. Somewhat counter-intuitively, to achieve a doubling of perceived volume in an audio system requires not double but ten times the power. That's because, as mentioned earlier, our ears work on a logarithmic scale.

TIP: To achieve a doubling of the perceived volume created by an audio system requires ten times the power, rather than the two times you might expect, due to our ears working on a logarithmic scale.

Ratios are all well and good, but when working with audio equipment we also need some real values to work with. While the decibel only deals in ratios, anything expressed in dBm is a fixed value, with 0dBm equating to 1 milliwatt of power dissipated into a 600 ohm load. You'll be pleased to hear that this is of little direct relevance in the world of modern audio, but was vitally important in the pioneering days of the telephone system when small amounts of electrical power needed to be transmitted over long distances. In the world of telephones and 600 ohm line impedances, 0dBm equates to a signal of 0.775 volts applied across a load of 600 ohms.

In modern audio applications we are far more concerned with the voltage than the load impedance so we tend to use the dBu (the 'u' meaning 'unterminated'): this signifies a signal level of 0.775 volts but without the load impedance being taken into account. In other words, while dBm is a measure of power, dBu is a measure only of voltage. The term dBv (with a lower-case 'v') is often used in the USA to mean the same thing as dBu. Of course, a reference voltage level of 0.775 volts is pretty clumsy, so some kind soul came up with the designation dBV to signify a signal level of 1 volt without regard to the load impedance. This was so mind-bogglingly sensible that most serious audio engineers declined to adopt it!

Somebody once said "the nice thing about standards is that there are so many of them to choose from", but fortunately, in analogue audio we tend to use mainly two standards for line level, with most equipment specified as working at either '+4dBu' or '–10dBV'. That's the level at which the meters read 0VU. Plus 4dBu is an operating level adopted in pro audio due to historic rather than purely logical reasons, and corresponds to an average signal level of 1.23 volts. This is a fairly convenient figure for use with modern circuitry as it leaves a sensible amount of headroom (typically about 20dB, in fact) before the circuitry runs into clipping.

Minus 10dBV corresponds to 0.316 volts and is used by some electronic instruments and consumer audio devices as well as some budget live sound and studio gear. The input gain range on a typical mixer will happily accommodate both standards, which differ by roughly 12dB.

Things are a little different in the digital world, as there is no headroom once the meter hits the end-stops signifying that Digital Full Scale has been reached. This means we should always work at an average level well below 0dBFS in order to create sufficient headroom—the loosely adopted 'standard' seems to be that –18 or –20dBFS roughly equates to working around the 0VU point in an analogue system.

Line-level signals are typically used by rackmount effects units, tape machines, MP3 players and many electronic keyboards, while mic-level signals, as the name implies, are what you'd expect to see from a typical microphone, although many active DI boxes are also designed to output mic-level signals to allow them to be connected directly to a mixer's mic input.

Balanced Audio

Balanced audio connections are used for microphone connections and many line connections in a bid to reduce susceptibility to interference that might otherwise cause hum or buzz. A typical balanced cable contains two wires twisted together, and encased in a conductive outer screen. The balanced input works differentially, which means that it only responds to the difference between the signals presented on the two signal wires; it ignores anything that appears identically on both (known as a 'common mode' signal). In order to reject unwanted interference that might find its way onto the two signal wires, it's essential that the interference induces identical voltages on both lines (to make it a common mode signal). That can only happen if both signal wires have identical impedances to ground, because voltages develop across impedances—so the 'balanced' part of the balanced interface refers exclusively to the balanced (i.e. matched) impedances to ground on both signal lines, and thus the unwanted interference is presented as a common mode signal which is ignored or rejected by the differential receiver.

In contrast, the wanted audio signal is presented as a differential signal so that it is recognised and passed on by the differential receiver. The old-school way of achieving that was with a transformer, in which case the signal on the 'cold' signal line has the same amplitude but opposite polarity to that on the 'hot' signal line. This is a 'symmetrical' balanced signal; both wires carry the same thing, but one is inverted with respect to the other. As it is the difference between these two signals that is passed on as the wanted signal, the output will have twice the amplitude of either one individually, and so typically the signal on each line is half the required amount, or -6dB. A lot of electronically balanced outputs emulate the transformer's symmetrical format, and thus work in exactly the same way. Typically there will be two active output drivers, one connected to the 'hot' signal wire, and the other to the 'cold' signal wire, the latter typically providing an inverted version of the signal from the former.

However, this isn't always the most practical solution in live-sound situations because if an electronically balanced output is connected to an unbalanced destination, only the signal on the 'hot' wire will be passed and often ends up being 6dB lower than it should be. A common alternative that solves this problem is the 'impedance balanced' output. This is badly named, since all balanced interfaces are inherently impedance balanced, but what is meant is that the wanted signal is only presented on the 'hot' wire, and a resistor is connected between the 'cold' wire and ground to ensure the correct impedance balance. With this configuration, the differential receiver at the balanced input sees signal on one wire and nothing at all on the other, but the difference between these is still the wanted signal, just as with the symmetrical approach—so it works in exactly the same way as any other balanced interface. The advantage, though, is that the hot signal is at the correct level to work as intended with unbalanced destinations.

TIP: Another commonly-used balancing system, usually described as 'ground-compensated' or 'impedance-balanced' uses the same wiring and works just as effectively—it is described in more detail in Chapter 13.

The key point is that noise currents induced into the screen of a balanced audio connection don't directly affect the audio signal as the screen does not form the audio return path, as it does in unbalanced systems—a benefit that makes ground-loop hum less likely.

For microphone signals and many line connection applications, the rugged, three-pin XLR is the standard connector, although many line-level connections are made via TRS (tip-ring-sleeve, three-pole) quarter-inch jacks. The Neutrik Combi socket used

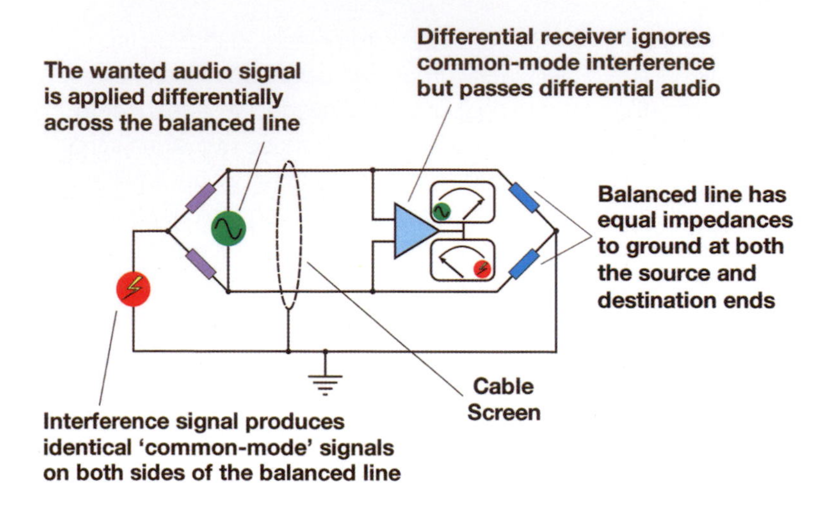

How a 'balanced' audio connection works to minimise noise pickup.

on some mixer inputs and outboard gear has a jack socket in the centre with XLR pin sockets around the edge so that it can accept either type of connector.

Phantom Power

Capacitor microphones have built-in electronics that require a power source and this is usually supplied via the mixer's 48 volt phantom power source. The term 'phantom power' was coined as the power source is invisible, being fed along the conductors of a standard balanced mic cable, so no additional wiring is necessary. The moving-coil (often called 'dynamic') microphones commonly favoured for live performance don't actually need phantom power to operate, but as long as they are connected via a correctly-wired balanced cable they won't be affected if the phantom power is switched on. This is just as well, as many smaller live-sound mixers have a global phantom-power switch that applies 48 volts to all the mic inputs. As a rule, if the mic body is fitted with a three-pin XLR socket, its output is balanced and it will be safe to use with phantom. Use of unbalanced cables can cause damage to dynamic mics when the phantom power is switched on, especially ribbon models, as can plugging or unplugging them with the phantom power turned on. Having said that, moving-coil, dynamic mics are pretty hard to kill!

TIP: Dynamic, moving-coil microphones, which don't need phantom power to operate, can safely be connected to phantom-powered inputs, so long as a correctly-wired balanced cable is used.

Channel Gain

The actual output from a microphone depends on both its electrical sensitivity, its proximity to the sound being picked up and the volume of that sound. Most mixer channels have a gain trim near the top of the channel that allows the mic signal to be adjusted to the optimum level before it continues its journey through the mixer. Line-level signals usually need to be attenuated slightly which is why mixers either have a mic/line switch or separate mic and line channel inputs.

In theory, to get the best sound quality, you should adjust the channel trim control so that the signal averages around the 0VU mark on the meters when the channel is faded up. This ensures that the signal level is well above the inherent noise floor of the mixer—even the best electronics produce a little residual hiss—but well below the overload level to allow for unexpected loud peaks or just the volume creeping up during a gig.

Many mixers have warning 'peak' LEDs on each channel to let you know when the signal level is too high to avoid the peaks of the waveforms being 'clipped' or squared off, which produces a very unpleasant-sounding distortion. On the other hand, if

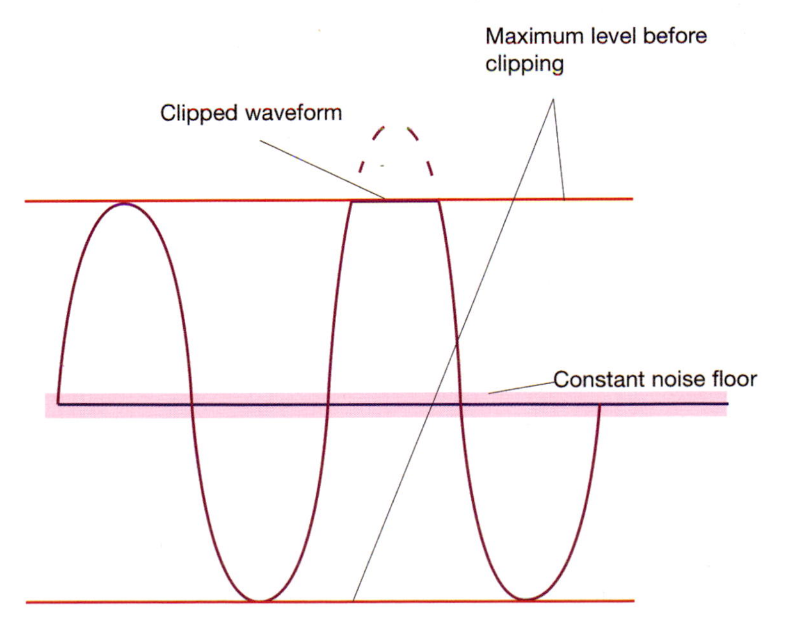

➤ For optimum sound quality the signal needs to be well above the noise floor but not so high as to allow any parts of the waveform to reach the clipping level.

they're way too low you'll need to push the channel faders further up and some background hiss may become audible. In extreme cases, if the input trim is set too low there may not be enough gain available on the channel fader to get the signal loud enough in the mix.

To further simplify setting the trim controls, most mixers have some form of 'pre-fade listen' (PFL) or Solo button that leaves the soloed channel audible in the phones output (and control room output if the desk also has studio applications) but mutes all others, allowing the selected channel's signal to be checked in isolation, while the main mix remains unaffected. (If several Solo buttons are pushed down at the same time, all the soloed channels are heard in the phones mix). These buttons are usually found close to the Channel On or Mute switch, generally somewhere close to the channel faders and although not part of the channel's input stage, they are important in setting the input gain trim correctly, which is why I have mentioned them early on.

PFL/Solo

PFL monitors the channel signal before the fader so that you can hear the channel signal even if the channel fader is turned right down. Solo or 'Solo-In-Place', on the other hand, monitors the channel signal after the fader (and pan control) so that the signal is heard at the same level it would be in the mix. On larger mixers you may also find a 'Solo Safe' function that allows certain channels to remain audible even when Solo is activated. For example, when you solo a channel you may still need to hear any effects being fed from it, so the effect returns would be set to Safe. Depending on the size and complexity of the mixer, Solo functions may also be included for any additional buses and for the effects sends and returns.

Other buttons you may find in the preamp section include 'polarity-invert' (often labelled Phase), as well as Pad and Low-cut. Polarity-invert is useful when multiple microphones are picking up the same sound source at differing distances, and can be used to minimise the resultant comb-filtering or phase-cancellation effects. These are often omitted on simpler mixers.

Solo and Mute buttons—on a live console Solo will usually only solo signals into the headphones or a dedicated output.

TIP: In practice, you can probably activate the low-cut filters on every input other than kick drum and bass instruments without risking anything sounding too thin.

Gain trim control, with polarity invert and low-cut filter below and phantom switch and mic/line selector above.

The Pad switch reduces the level of the incoming signal, usually by 15 or 20dB, which is useful when you have a highly sensitive microphone placed close to a very loud sound source, such as a kick drum or guitar amplifier. Without the Pad switched in, the microphone output might be high enough to overload the channel's input stage, causing distortion.

The Low-cut switch inserts a filter to remove unwanted low frequencies below about 80Hz—such as vocal popping or stage rumble. As a rule, low-cut filters can be switched in for just about any type of input other than kick drum and bass instruments.

Channel EQ

After the input preamp and its gain trim comes the equalisation section, which can be as simple as just bass and treble

This Yamaha mixer's channel EQ includes shelving high and low bands, with two sweepable mid bands.

(low- and high-frequency shelving controls), although all but the smallest mixers now tend to have one or even two midrange EQ controls as well. A typical EQ range is 15dB of cut or boost. On some mixers, the midrange section offers variable amounts of cut or boost at fixed frequencies, but more sophisticated models may offer a 'sweep' EQ which allows the cut or boosted frequency to be adjusted. Another EQ parameter that you may encounter is midrange 'bandwidth' or 'Q', either as two switched settings, or a variable control which adjusts the range of frequencies affected. The EQ gain controls are usually detented so that you can feel a click when they are in their neutral position, but ideally the EQ will be bypassable to make it easier to compare the orginal and equalised signal.

Pre-Fade Auxiliary Sends

Auxiliary sends are used to create separate mixes to feed foldback/monitor speakers or effects, and there are two types, designated pre-fade and post-fade. A pre-fade send is essentially a level control sourced from directly after the EQ stage, feeding a mono mix bus which combines the send signal from each channel. The signal is then fed via an Aux Master level control to an Aux output. Pre-fade sends are usually used to feed stage monitors. Smaller mixers may have only one pre-fade send whereas a large console will have several, allowing a number of different monitor mixes to be created. As it is sourced before the channel fader, the pre-fade aux signal level is completely independent of the channel fader position, which means that the monitor-mix balance won't change when the front-of-house balance is adjusted.

Post-Fade Auxiliary Sends

The post-fade send controls may be located right after the pre-fade sends on the control panel, but they take their feed from after the channel fader. The send level is affected by changes in the channel fader position, which is exactly what we need when feeding effects, such as reverb or delay. This way, when the channel signal level is turned up or down using the fader, the amount of added effect changes by the same amount so that the proportion of added effect remains constant. If you were to feed a reverb effect from a pre-fade send, the reverb level would stay constant, regardless of the fader setting and in the most extreme case the reverb would still be audible with the channel fader turned down to zero.

By using different settings of the post-fade control on each channel, it is possible to send different amounts of each channel's signal to the same effects unit, the output of which is usually mixed into the main stereo mix. In other words, one reverb unit connected via a post-fade aux output allows different amounts of reverb to be added to each channel.

Note that an effects unit fed from a post-fade send should be configured to output only the processed sound and none of the original input signal, which is achieved by setting the effects mix control or parameter to 100% effect, sometimes called 'wet'. The 'dry' or direct component of the sound comes via the mixer channel. As with pre-fade sends, most mixers have multiple post-fade sends, allowing several different effects to be used at the same time, all of which can be applied independently to the different channel signals.

Pre- and post-fader aux sends perform different functions, so most desks will have some of each, or the option to switch between modes.

Fader, Pan and Balance

The channel fader, which may be a rotary knob on some smaller mixers, comes directly before the pan control, and is usually accompanied by an On or Mute button. Where sub-group routing buttons are provided these are traditionally located close to the channel fader. On an eight-bus mixer, the routing buttons would be marked 1–2, 3–4, 5–6, 7–8, with a further L-R button for routing the channel directly to the stereo mix. A pan control then adjusts the balance of the channel signal between the master left and right mix bus (or between odd and even bus numbers if the mixer offers routing to other buses).

When a pan control is turned one way the channel signal is routed only to the left mix bus while turning it the other way routes it only to the right bus. Leaving it centred sends equal

Routing buttons in the input channel path will send the signal straight to the main output, or via a sub-group bus.

amounts of signal to the left and right mix buses placing the sound centrally in the mix—assuming you have a stereo speaker system connected. Where a mixer has dedicated stereo channels, as many of the smaller ones do (usually line-level only), the pan control is replaced by a stereo balance control that changes the balance of the left and right stereo signals passing through the channel.

The left and right mix buses are often referred to as a single stereo mix bus and a master level fader or knob controls the final output level from the mixer, allowing it to present the correct signal level to the speaker amplifier.

Insert Points

Insert points offer another means to connect an effects unit or signal processor. All but the smallest mixers tend to have at least some insert points, often in the form of unbalanced TRS (send/return) jack sockets. Essentially, an insert point allows the normal channel, bus or stereo master signal path to be interrupted, routed through an external device and returned to the channel signal path. Channel insert points usually come after the channel's input stage but before the EQ section, although on some models the insert point may be switchable pre- or post-EQ. When nothing is plugged into the insert point, spring contacts inside the jack socket close and allow the signal to flow through the channel uninterrupted.

Note that although the insert point uses the same socket as used for stereo and balanced jacks, the usual protocol is that the screen provides a common ground while the two inner conductors provide an unbalanced send and return to the external device. To make use of an insert point you will require a specially wired 'Y cable' with a TRS jack at one end and a pair of mono or TS jacks at the other end for the send output and return input connections. Most mixer's TRS sockets are wired 'Tip-Send / Ring-Return', although not all of them follow exactly the same protocol, so you may need to swap the two mono jacks over if it doesn't work first time.

Multi-Bus Mixers

In a more complex live-sound setup, it is often convenient to be able to control groups of related sounds, such as all the mics over a drum kit, using a single fader. This is where those extra mix-buses or 'sub-groups' come in handy. The number of sub-groups varies depending on the model of mixer and buses can be used either in mono or in pairs to carry stereo signals.

Any of the input channels may be routed to any of the available buses using the routing buttons, with the channel faders setting the relative levels of the various elements in the mix while the corresponding bus fader, usually located in the mixer's master section, controls the overall level of any channels routed to that bus. The channel Pan control determines whether the signal is sent to the odd or even numbered bus as the routing buttons almost always select pairs. With the Pan control central both buses receive the same signal.

The outputs from these bus 'groups' are usually routed into the stereo mix. So, for example, to change the drum-kit level you

Routing to sub-groups allow individual sections of the mix to be controlled from a single fader.

simply use the sub-group fader(s) to which they are routed instead of having to adjust six or more input channel faders. In addition to a drum submix, you may want to set up further groups for backing vocals, keyboards and so on, depending on how many buses your mixer has and whether you need to use them individually in mono or as stereo pairs.

Built-in Effects

A number of small and medium-format live sound analogue mixers come complete with a built-in effects section, usually offering one or two effects at the same time fed from the mixer's pre-fade sends. Sometimes it is possible to bypass the internal effects and connect your own external effects boxes if preferred. The feature set of these inbuilt effects range from having presets that can be varied only in the amount of the applied effect, all the way up to having full control over the most important parameters, such as reverb time, reverb pre-delay, delay time and delay feedback. Perhaps the most common problem associated with the preset type is that far too many of the presets either have ludicrously long reverb times or are given over to gimmicky but largely pointless effects such as distortion, telephone simulation, robot voices and so on. For most PA applications, a good plate reverb emulation and an adjustable delay/echo will do just fine. If there's a button that lets you tap in the tempo of any delay effects, so much the better.

Master Section

A combined stereo (or two mono) output fader control the output level of the main stereo mix bus. Output levels are generally monitored via a meter, most often a bar-graph type, which normally also doubles as a PFL monitor, to help set the input gain for individual channels. There's usually also a PFL warning light in the master section to let you know that one or more Solo/PFL buttons are active. A stereo mixer with no additional sub-groups is described as having a 'something into two' format, with the 'something' being the number of input channels. For example, an eight channel mixer with a single stereo output would be an eight-into-two, or 8:2 mixer.

In addition to the main output fader, the master section might also include any additional sub-group faders, mono/stereo switching for the buses, aux send masters and effect-return channels. There may also be a separate 'control room' section—usually found on dual-purpose stage/studio models—plus a headphone output that allows the operator to monitor the PFL signals without affecting what the audience hears. Any inbuilt effects will also be selected in the master section.

The output from an external effects unit may be connected into the mixer via spare input channels or via dedicated effects return inputs, also known as aux returns—these are electrically similar to input channels but usually offer more limited facilities, such as no mic inputs and often no EQ. Normally, effects returns feed straight into the main stereo mix, although larger

Inputs, sub-groups and effects returns all end up at the master faders.

mixers may include routing buttons and sends allowing them to be added to monitor mixes. A spare input channel or pair panned hard left and right for stereo, can always be used as an effect return, offering the benefit of EQ and access to the aux send buses (to add reverb to the cue mix, for example). However, make sure that the corresponding aux send is turned down on the return channel, otherwise the effect signal will be fed back on itself, resulting in a feedback loop.

Connecting Effects And Processors

Understanding how different types of outboard can and can't be connected will save you a lot of trouble and frustration later on. While it is okay to connect any type of effect or signal processor via an insert point, there are some common-sense limitations as to what can sensibly be used via the aux send/return system.

As a very general rule, only delay-based effects such as reverb, echo and modulation effects should be connected via the aux system—I like to call these devices 'effects', as opposed to 'processors'. If the box uses delay to do its work, it's almost certainly an effect, even if the amount of delay is very small, as in the case of chorus and flanging. If there's a 'dry/effect' mix knob or parameter, the box is also likely to be an effect. The key point to appreciate about a typical effect is that it is intended to be added to the original signal. When an effect is connected via an insert point, its wet/dry balance is set using the mix control on the effects device itself. Conversely, when the effect is being fed from an aux send, the dry part of the signal is already being provided via the mixer channel, so the mix control on the effects unit or plug-in should be set to 100% wet, 0% dry.

TIP: Be sure to use compression sparingly at smaller gigs, at least on microphone channels, as any gain reduction applied by the compressor means you lose some of your safety margin before feedback occurs.

A process such as EQ or compression doesn't add to the original signal, it changes it. The processor family includes compressors, limiters, noise gates and EQ, and so would normally be used only via insert points, not via the aux sends and returns. Inevitably, there are exceptions, such as when setting up the parallel compression effect popular in recording studios—in this case a send is used to add a very heavily compressed version of a signal to itself in order to beef it up.

Having said that, I'm always very wary of using compression in any form for smaller gigs, at least on microphone channels, as any gain reduction applied by the compressor means you lose some of your safety margin before feedback occurs.

Powered Mixers

Racks of separate amplifiers are bulky and require extra cabling, which is why active PA loudspeakers have become so popular. Another popular choice for simple systems is to use a mixer with power amplifiers built in, combined with passive speakers. 'Powered mixers' used to be rather heavy, but the advent of Class-D amplifier technology and switch-mode power supplies has made them considerably lighter. Of course, this does leave you slightly vulnerable if the amplifiers in the mixer fail.

Another option offered by some powered mixers is to switch the internal power amplifiers to drive passive monitor speakers, leaving the main outputs to feed active speakers just as you

Modern lightweight power amps can now be built into mixers, forming a neat self-contained option for those who prefer to use passive loudspeakers.

would with a non-powered mixer. Working this way gives you some redundancy in case of a system failure, but if you're worried that a mixer amp failure may prevent you using the mixer section as well as its internal amplifiers, you could take a little mini-mixer in your gig-bag as backup just for the vocals. These are very cheap and can save the day when all else fails.

Digital Live-Sound Mixers

Large-format digital consoles have been used very successfully in live sound for a number of years now, but for the gigging musician, a simple analogue desk, ideally with on-board effects, is often still the preferred option because of its straightforward layout and absence of 'hidden' features. However, that is gradually changing, with a number of affordable, small-scale digital consoles appearing on the market in recent years. To answer the question as to whether you'd be better off switching to a digital mixer or not, it's necessary to scrutinise the essential differences between a typical digital and analogue mixer.

Digital mixers can offer a lot of mixing power in a very small package.

Personally, I don't think that audio quality is a significant factor these days, as both types are capable of producing excellent results, even at the lower-cost end of the market. A digital mixer will still have analogue mic preamps, and they play a large part in determining the quality of sound, but these days it is rare to come across a truly poor mic preamp even on budget equipment.

So, is a digital mixer more reliable? Both types tend to be pretty robust, although if a small digital console fails, the chances are that the whole thing will stop working whereas with an analogue mixer you may just lose a channel or a block of channels depending on the nature of the fault.

Going GUI

Often, a digital mixer will be smaller than its analogue counterpart, as the user interface is usually stripped down so that instead of getting 'one-knob-per-function', as you would on a typical analogue desk, you'll often find fewer assignable physical controls and they change function depending on context. For example, there may be just a single set of controls for the EQ and aux sends that apply to whichever channel is currently selected, and there may also be more channels than there are physical faders, with a 'bank' selection system used to allow the faders to operate the extra channels. Sometimes, the same faders may also be used to address the aux sends, which is particularly helpful when setting up monitor mixes.

Although there are many common elements between mixers from different manufacturers, there is no standard paradigm for digital desks, so you have to decide what compromises are acceptable to you when juggling features, price, control layout and number of channels. You will also find that on some desks the faders are motorised, allowing them to follow the value of any recalled settings, whereas with other designs the faders are entirely manual, so their physical position will not match the actual settings once you recall a new scene from memory. Some may even have no faders at all, being controlled entirely from a touch screen instead—but at least the virtual faders will always move to show the correct value. Rotary controls are less of a problem as the manufacturer can use rotary encoders that turn continuously with a ring of LEDs around each encoder to denote the current value.

Recall

One of the major advantages of digital desks is that they are very good at remembering complete sets of control settings for instant recall (the only parameters usually excluded are the analogue mic trims and master output level). These recalled control settings are usually known as snapshots or scenes, and once stored, the user can recall any one at the press of a button: all the fader settings, the EQ, the send levels, any on-board effect and processor settings, and all the monitor mix settings. This ability to recall and change snapshots is clearly invaluable where the same show is being repeated, such as when playing several nights at the same venue or putting on a theatre production. Being able to save EQ and effects settings for individual performers is useful, even if you have to tweak these a little to suit the venue. With most mixers you can also load new effects settings without affecting the rest of the scene. Another advantage of digital mixers is that their control software can often be updated, so your mixer might actually improve with age!

The Right to Roam

Many of the newer small-scale digital mixers have another very valuable trick up their virtual sleeves—wireless remote control. How this is incorporated varies from model to model, but the general paradigm is that you use some sort of computer device, such as a wi-fi-equipped laptop or tablet, to operate the mixer from anywhere within wi-fi range. Some have this facility as an extension of their hardware interface, while others use it as an alternative.

Big concert system users have enjoyed this way of working for some time, allowing the engineer to walk around the venue and check for coverage in all the different seating areas, and in the land of the Dog & Duck pub, being mobile offers a number of very practical benefits, too. Not least is that you can simplify your stage wiring by leaving the mixer hardware on stage and plug the mics and DI boxes directly into it—no more stagebox and heavy multicore—and you can then use your computer or tablet in the normal mixing position, half or maybe two-thirds of the way back into the venue. This, for me, is a really big deal: multicores are heavy, they need a large reel for storage, they are a Health-and-Safety trip hazard and they have to be connected up at the mixer end, which means more plugs and sockets to manage. They also get trodden on and, at outdoor events, driven over, which can often lead to premature failure.

To give you a picture of how a typical small digital mixer works in practice, I'll quickly outline my experiences so far with Mackie's DL1604 digital mixer, which seems well-suited to any small venue where its 16 input channels and six monitor mixes are sufficient. Similar devices from other manufacturers are available.

The control device for the DL1604 is an Apple iPad running a free App that Mackie seems to update and refine as user feedback comes in. There are no physical controls on the mixer other than the input gain trims, although the iPad can be docked directly into the mixer where an all-in-one solution makes more sense, such as when you're mixing your own gig from the stage.

For remote use, a standard wi-fi router connects to the mixer via an Ethernet cable and your iPad connects via wi-fi. In larger venues, where you might be worried about wi-fi range, the router can be positioned elsewhere in the room and connected to the stage via an Ethernet cable, which is a whole lot easier to run than a multicore! Note that no audio passes through the iPad—it is used purely as a control surface, although when docked, it can also be used to play songs from iTunes and the like through the mixer.

So far I've had no problems at all with interference from other wi-fi devices, even at a festival where hundreds of people might have had wi-fi-enabled smart-phones. If you should lose the wi-fi link for any reason, the mixer carries on working but with all the settings frozen as they were when communication was lost. Control is regained seamlessly when the wi-fi link is restored.

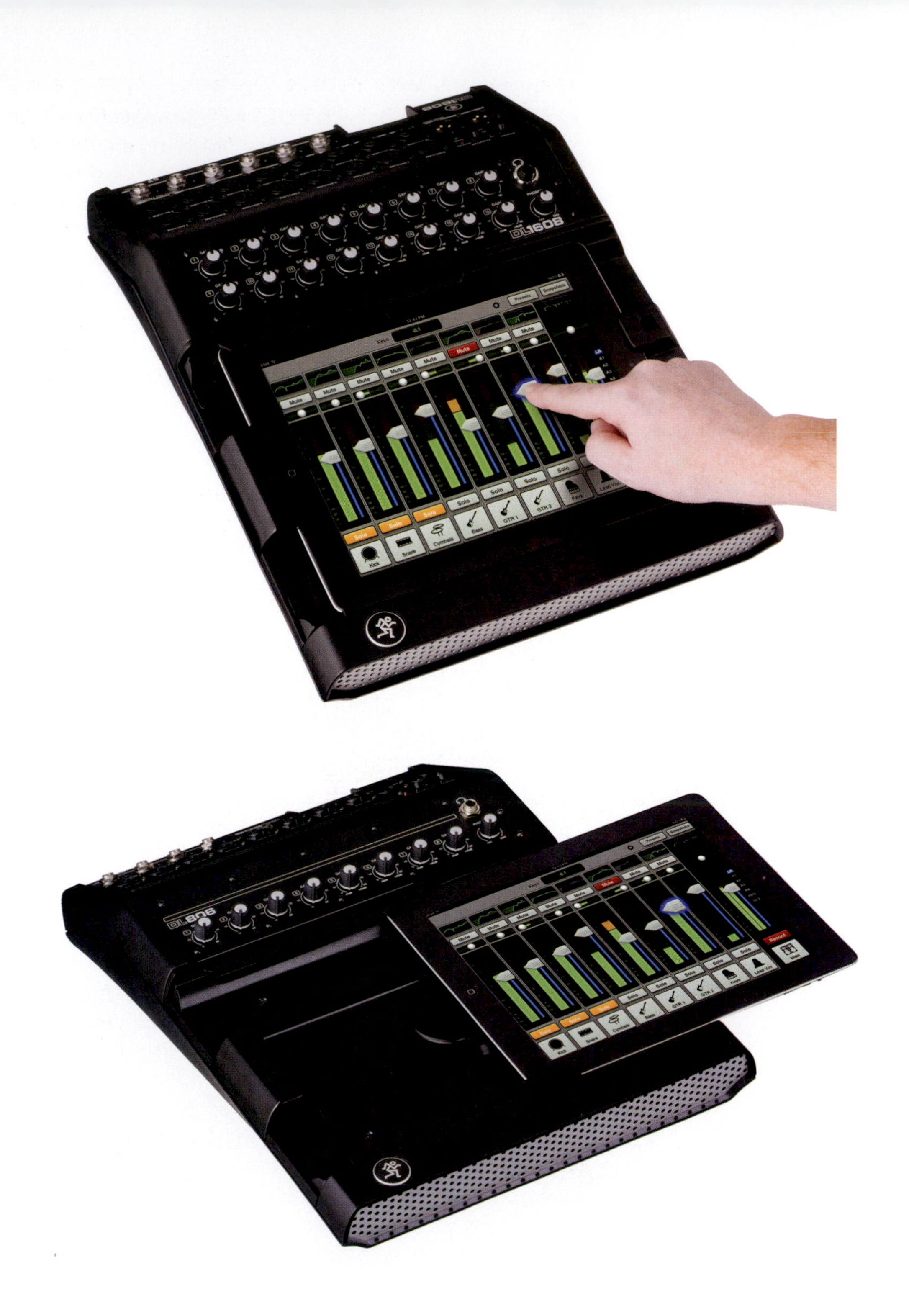

Mackie's DL1608 compact digital mixer employs an iPad as the controller which can be docked to the hardware mixer, or detached, allowing you to operate the system remotely via a wireless network link.

Mixing from a touch surface actually feels pretty good if you're used to using tablets, but it does mean you have to look at the screen when adjusting a fader, whereas with hardware faders you might, for example, keep your hand on the lead guitar fader and then watch the band to see when it needs to be turned up for a solo. All 16 channels are viewable across two screen widths, with the on-board delay- and reverb-return faders added to the end of the mixer. These same faders control the monitor levels using the 'layers' approach adopted by many digital mixers. There's also provision to allow individual performers to use their own iPhones or iPads to adjust a specific range of unlocked features, allowing them access only to their own monitor mix, for example.

Although the Mackie mixer is the only one I've used for any length of time, the approach taken by other makers is broadly similar. For example, the Line 6 StageScape mixer uses a wi-fi USB dongle rather than a separate router, as do the newer

Controlling the mixer remotely allows you to check the sound at different points in the venue.

Presonus mixers, while the older Presonus StudioLive models connect via a laptop and cable, allowing you to use a suitable Smartphone or tablet to talk to the mixer via the laptop. Presonus also have a very neat and simple Smartphone app that performers can use to adjust the balance of themselves relative to everything else in their own monitor.

TIP: One useful tip I can pass on here is that if you've been over-generous with the headroom on one channel and it isn't practical to go up to the stage and tweak the gain trim, you can insert a compressor onto the offending track with the ratio set at 1:1 so no compression is being applied, and then use the make-up gain slider to add more gain.

Not only does the wireless approach give you plenty of freedom when it comes to choosing a position from which to mix, it also means you can walk onto the stage during the soundcheck and hear how the monitor mixes sound first-hand while making adjustments. The only real downside of on-screen control is that for outdoor gigs on sunny days a computer or tablet screen can be hard to see unless you can arrange some shade.

Other Stuff

As long as the mixer has enough DSP power, the designers can add things that in the analogue world might require racks of outboard gear. For example, you might have access to compressors and gates on every channel and bus, plus comprehensive effects and third-octave graphic equalisers available on every output. There's also no reason models couldn't be extended to offer anti-feedback systems and spectrum analysers as DSPs get more powerful every year. Not every mixer will give you every feature, but the point is that much more can be included than you'd find in a typical analogue desk and the controls for these extra processors often follow the DAW plug-in paradigm, making them very easy to adjust.

On my very first major PA job using the Mackie DL1604, the band leader asked me to notch out very specific frequencies for each of four violins, as he knew the precise characteristics of the instruments and their pickup systems. If I'd taken my little analogue desk, that would have been a case of "You can have bass, middle or treble, sir . . . ", but using the fully parametric equalisers in the DL1604, I could give him exactly what he wanted.

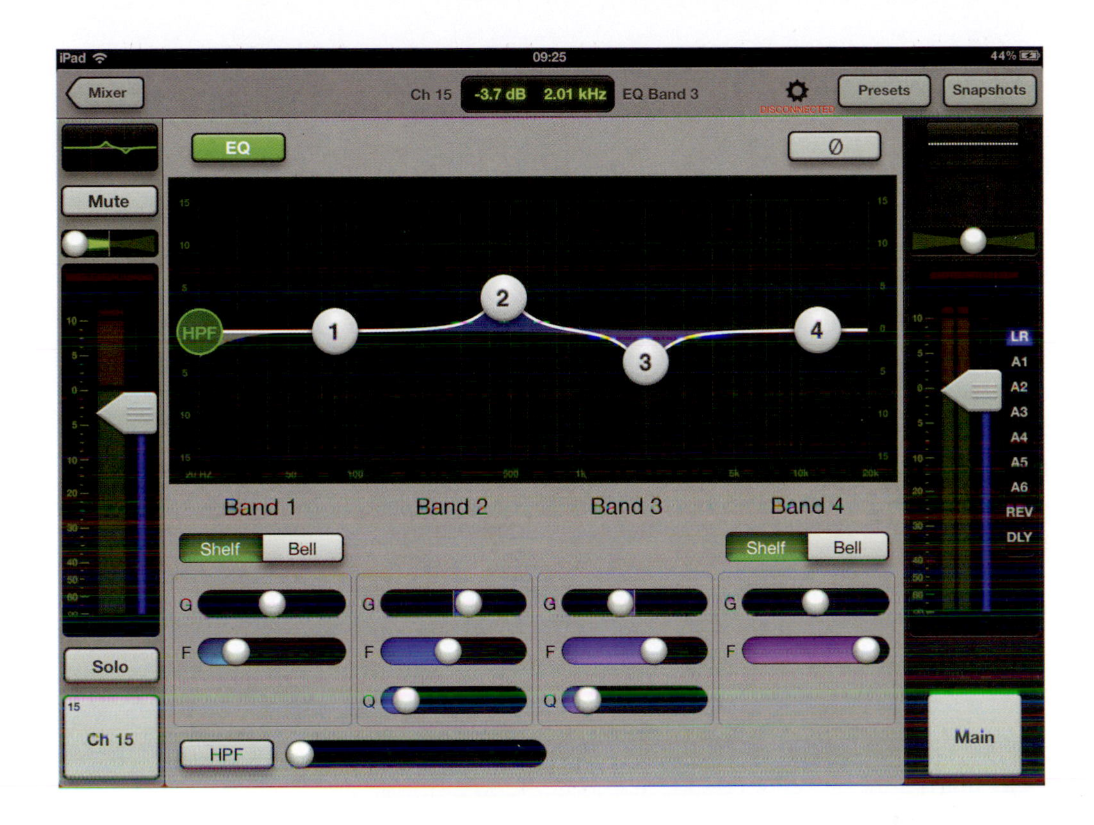

Digital mixers will generally feature more sophisticated EQ options than a compact analogue desk can offer.

DO YOU NEED A DIGITAL MIXER?

If you always mix from on-stage and you only need a bit of delay or reverb, then a small analogue mixer may still be the most straightforward option, but as soon as you find yourself in a situation where you need an engineer to mix your set, then a wireless digital solution starts to make a lot more sense. If nothing else you'll save a lot of time setting up and you won't need to carry a heavy multicore.

There's also the factor that even when the mixer and its control surface is on stage, having access to more detailed EQ as well as the option of third-octave graphic equalisers on all the sends and main outputs can be a big plus, as it saves carrying outboard hardware and the associated wiring clutter. You may also opt to run wireless with a tablet fixed to your mic stand using one of the many adaptors now available so that you can get to the fader levels and effect mutes without moving from your playing position.

Operational Differences

Digital mixers have slightly different operational constraints from analogue desks, but the same general philosophy applies insofar as the signal levels should be kept reasonably high while still leaving a practical headroom margin, to allow for unexpected signal peaks. The main difference is that when an analogue mixer is pushed into the 'red' on the meters there's plenty of headroom left so the distortion increases progressively beyond this point until you eventually run out of headroom and the thing clips. With digital audio systems, there's no significant distortion until you run out of headroom at which point the signal suddenly clips, usually sounding pretty nasty.

This needn't be a problem, in practice, as modern digital mixers have the same dynamic range as analogue mixers so it is possible to leave around 18dB of headroom without compromising the audio quality on low-level signals—the same as in analogue equipment when the signal level exceeds 0VU on the meters.

Chapter Five Microphones

The appropriate choice of microphone is as important in live sound as it is in the studio, and although most gigging bands get by using mainly robust, moving-coil dynamic microphones, it pays to at least be aware of the alternatives. Regardless of the operating principle, a microphone's purpose in life is to convert sound into an electrical signal that is compatible with a typical mixer's microphone input. Sound consists of complex, cyclic changes in air pressure, and most microphones are designed to detect these pressure changes using a lightweight diaphragm that can be easily moved by the air. Some form of electrical mechanism, such as a coil of wire in a magnetic field, is then used to translate the diaphragm's movement into an electrical signal. Obviously, the more accurately the diaphragm can track the air pressure variations, the more accurate the microphone's output can be made, so the ideal mic diaphragm assembly will be light enough to respond very quickly, to accurately pick up high-frequency sounds, while still being robust enough to withstand big changes in pressure generated by loud low-frequency sounds. The part of the microphone containing the diaphragm is known as 'the capsule'.

Pickup Patterns

In live sound we tend to gravitate towards directional microphones with a view to reducing spill and avoiding feedback, but there are occasions when other polar patterns may be useful. The simplest microphone, created by fitting a thin diaphragm across the mouth of a sealed cavity, picks up sound equally from all directions, and we call that an

The basic shape of hand-held live-performance mics doesn't tend to change much from one manufacturer to another as it is largely dictated by function.

omnidirectional (often shortened to just omni) mic. Another basic pickup pattern is the so-called figure-of-eight, which picks up from both front and rear, but not from the sides. This is achieved by having the diaphragm exposed to the air on both sides. By contriving to combine both of these pickup patterns we can create the unidirectional or cardioid (meaning heart shaped) response with which we are all very familiar. By changing the contribution of the omni and figure-of-eight patterns, the resulting cardioid pattern can be made narrower or wider.

The sensitivity of a microphone at different frequencies and angles can be plotted on a graph called the 'polar pattern'. The resulting lines should ideally reveal the intended pickup pattern, although the typical cardioid response tends to become more omnidirectional at low frequencies and slightly more directional at high frequencies. While all this may currently be of less concern to you than "When can I play my next guitar solo?", it's well worth being aware that the consistency, or otherwise, of a microphone's polar pattern at different frequencies can have a major influence on its susceptibility to feedback.

Omnidirectional

My description above of an omni mic was oversimplified: if a diaphragm is fixed across the end of a sealed, airtight cavity, it responds not only to sound but also to those much slower changes in air pressure caused by weather, so, effectively, we have a microphone that doubles as barometer! To prevent the diaphragm being pulled in or pushed out by slow changes in atmospheric pressure, the cavity behind the diaphragm is vented to introduce what is essentially a slow leak, which means that atmospheric pressure changes no longer have an effect. At audio frequencies, however, we can consider the cavity to be

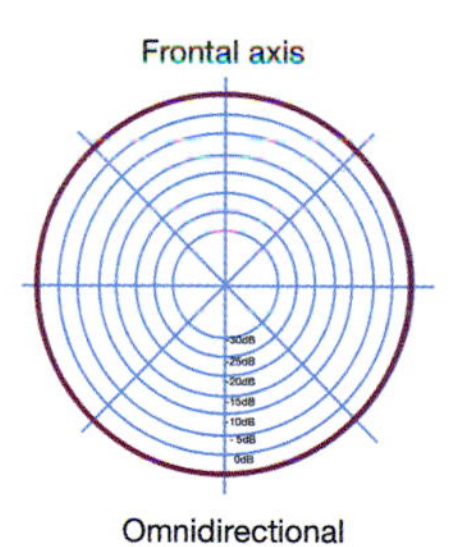

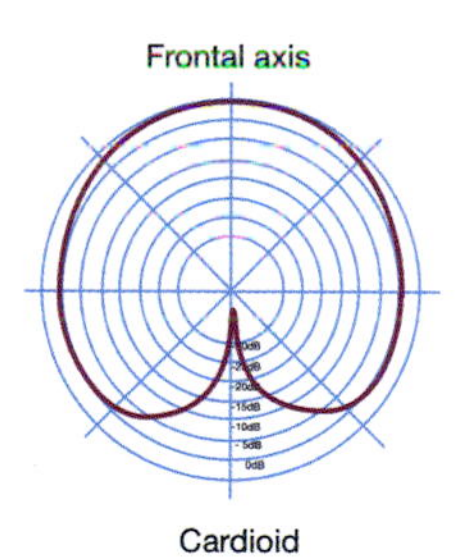

Frontal axis

Figure-of-eight

Frontal axis

Hyper-cardioid

Microphone polar response plots.

sealed. This kind of microphone senses absolute air pressure, known as 'Pressure Operation'.

Key points to note: the simple design of the omni capsule results in a very natural sound quality, but as omni microphones are not directional, they pick up more spill from other sound sources that lie off their frontal axis than do more directional microphones. Nevertheless, the 'quality' of the spill is generally very good as the frequency response is almost flat all the way round. Furthermore, omni mics don't exhibit proximity effect—the bass boost that you hear when a cardioid or figure-of-eight microphone is placed very close to a sound source—so you may often be able to use them with a closer positioning than a cardioid, reducing the ratio of spill to wanted direct sound.

Figure-Of-Eight

Although the figure-of-eight microphone is rarely used in small-scale live-sound applications, an understanding of its operation is useful, as the cardioid mic that is the mainstay of live sound is essentially a hybrid between an omnidirectional and a figure-of-eight polar pattern. The figure-of-eight microphone is based on a very simple configuration: its diaphragm is open to the air on both sides, and as any general increase or decrease in atmospheric pressure will affect both sides of the diaphragm equally, there will be no movement. Similarly, sound arriving from the side (90 degrees off-axis) of the diaphragm reaches both sides simultaneously, so there will be no movement and therefore no electrical output.

Rather than measure absolute pressure, the figure-of-eight mic responds only to a difference in pressure between the front and the rear of the diaphragm. This arrangement gives us what is termed a 'pressure gradient' microphone. If sound approaches the diaphragm on-axis from either side, an electrical output results, but the sensitivity reduces as the source moves further off-axis, as the pressure difference between front and back reduces, until at 90 degrees the microphone is effectively 'deaf'—from the shape of the polar pattern, it is immediately obvious how the figure-of-eight pattern got its name. Positive air pressure on one side pushes the diaphragm in one direction while positive pressure on the other side pushes it in the opposite direction, so the electrical output has opposite polarities for front and back sound sources.

Ribbon microphones are not the most robust as they use a delicate conductive ribbon suspended in a strong magnetic field. By contrast, the more common moving-coil mics, such as the ubiquitous Shure SM58, operate using a diaphragm fixed to

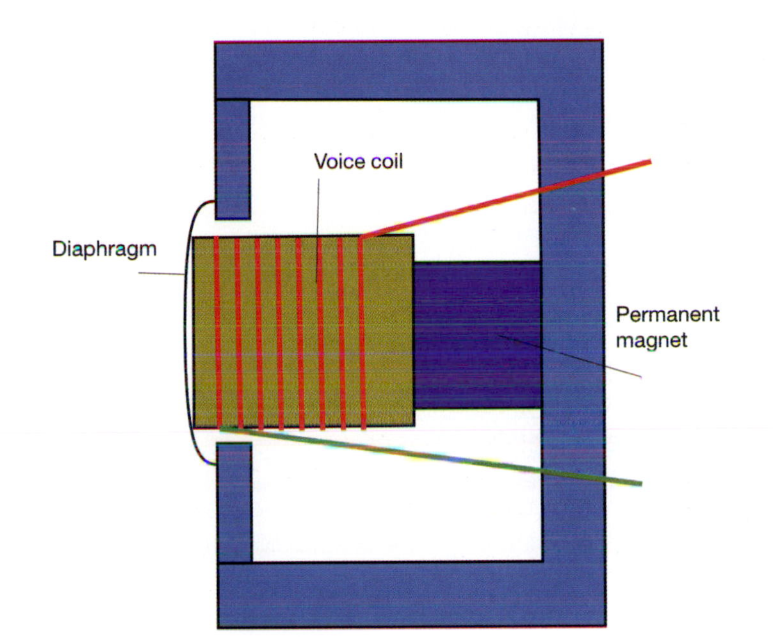

Dynamic microphone capsule.

a cylindrical former onto which is wound a coil of thin, insulated wire. This voice coil is suspended in a magnetic field and as sound causes it to move, an electrical output is generated, operating a little like a loudspeaker in reverse.

Pressure-gradient microphones exhibit what is known as 'proximity effect'—an increase in the mic's sensitivity to low frequency sounds when it is used close to the source. The figure-of-eight microphone has the most pronounced proximity effect of all microphone types, with a bass lift that can be up to 20dB.

Cardioid

The unidirectional or cardioid mic employs a specially designed sound labyrinth to delay sounds reaching the rear of the diaphragm, which produces the familiar heart or apple-shape polar pattern.

If you were to overlay omni and figure-of-eight polar plots, then add all the points together taking note of the opposite polarity rear figure-of-eight lobe, you'd get a standard cardioid polar pattern, with the mic still being fairly sensitive to the sides and a dead spot directly behind the capsule. If the figure-of-eight contribution is increased, a small lobe starts to appear at the rear of the mic with dead spots either side, usually 30 to 45 degrees off the rearward axis. This creates hyper-cardioid or super-cardioid mics, and their frontal pickup pattern is narrower than that of a cardioid. A key point to appreciate however is that the off-axis frequency response directional microphones can be far from even, so although it may pick up a little less spill than an omni, the 'quality' of the spill tends to be much worse, often sounding coloured or dull. Cardioid mics have a less pronounced proximity effect than figure-of-eight mics, but it can still be very noticeable.

TIP: Although the polar response plots of microphones are shown as two dimensional graphs, you must keep in mind that the actual pattern is three dimensional, so an omni mic's coverage pattern is a sphere, not a circle. This is important in the context of cardioid and hyper-cardioid mics, as knowing where the dead zones are in 3D will help you position your monitors correctly to minimise the risk of acoustic feedback.

Which Pattern?

Cardioid and hyper-cardioid mics are preferred in live sound, to minimise unwanted pickup from other instruments and to avoid feedback. However, omni mics shouldn't be completely discounted as, in situations where these can be placed close enough to a sound source that level before feedback isn't an issue, they will generally produce a more accurate, 'natural' sound. For example, when close-miking a drum kit—a situation where all the mics inevitably pick up significant spill from the other drums—the use of omnis may well give you a cleaner, clearer overall sound. It is worth remembering that an omni mic positioned at between half and two thirds the distance from the sound source as a cardioid mic will result in a similar amount of spill, so omni mics are often perfectly usable in situations where your instinct might suggest a cardioid would be the only option.

Figure-of-eight microphones are rarely used on stage, although some performers use one to recreate a vintage 'crooner era' sound. The main problem is that ribbons are fragile and dropping them, or even exposing them to strong blasts of air, can stretch or break the extremely thin metal diaphragm.

Other Considerations

The mesh grilles of live vocal mics offer some resistance to popping, caused when loudly sung 'plosives'—P or B sounds—send a strong blast of air onto the diaphragm. However, the mesh is usually too close to the capsule to be 100% effective, so some mic technique on behalf of the singer is still helpful, for example, turning the head slightly to one side when pronouncing these plosives.

Different mics also exhibit different degrees of handling noise, so hand-held performance also requires a degree of technique, for example holding a loop of cable with the mic not only takes some stress off the connector but can reduce handling noise. Also, and very importantly, vocalists need to be aware of the mic position relative to the monitors and to avoid holding the mic pointing down towards the monitors during instrumental solos. An awareness of proximity effect is also important, as it can be used as a performance tool by the vocalist who may vary the mic distance to create a more intimate sound or to vary the performance dynamics.

Moving-Coil Dynamic Mics

Many small gigging-PA setups employ solely moving coil dynamic microphones for vocals, backline amplification and drums, and these will almost always have a cardioid pickup pattern. As live vocal performance invariably means working close to the microphone, the majority of dedicated vocal mics exhibit a built-in low-frequency roll-off to help counter the proximity effect that would otherwise occur.

Dynamic microphones are relatively inexpensive, need no external power, and they are the most robust of all microphone types, being reasonably tolerant of knocks, smoke, breath condensation and high SPLs. On the negative side, the mechanical inertia of the moving parts restricts the capsule's efficiency at higher frequencies—a typical dynamic microphone will work effectively up to around 16kHz before the response begins to roll off. However, in practice, a good dynamic microphone sounds absolutely fine for live vocals, drums and most instruments.

Dynamic microphones produce a relatively small signal for a given level of sound, but for live use where microphones tend to be used close to the sound source, this is not a problem—the mixer's mic preamp will have more than enough gain range to accommodate close miked vocals, guitar cabinets and drums.

Capacitor Microphones

The capacitor microphone works on a very different principle from the dynamic microphone and has a number of technical advantages. Capacitor mics use an electrically conductive diaphragm held close to a fixed backplate and as sound moves the diaphragm this generates a change in capacitance. This is converted into an electrical output by means of some fairly simple active circuitry. With no voice coil attached to it, the diaphragm can be made of a very lightweight material, so it can respond much more accurately to high frequencies, with a response extending to 20kHz and beyond.

The diaphragm of a capacitor mic needs to carry an electrical charge induced by a high polarising voltage, and the active

Capacitor microphone capsule.

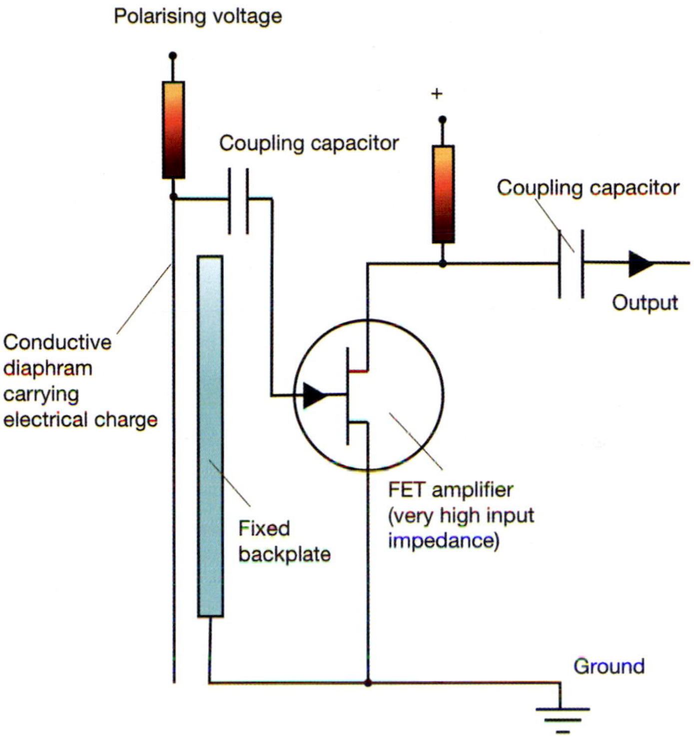

circuitry requires power, too, and both are usually provided by the 48v 'phantom' power source built-in to most mixing console or mic preamps.

Electret Capacitor Microphones

While a traditional DC-biased capacitor microphone requires a power supply to charge the capsule, electret microphones use a permanently charged 'electret' material within the capsule itself. This electret material is usually attached to the backplate, creating the popular 'back-electret' microphone. As a high polarising voltage is no longer required, some of these microphones can operate either from phantom power or batteries. Numerous hand-held capacitor and back-electret models have been produced for vocal use, the aim being to rival the sound quality of studio microphones in a live performance context. Miniature back-electret designs are also commonly used for clip-on drum mics.

A well-designed capacitor mic offers low noise, high sensitivity and a wide frequency response. They can also be designed with any of the popular polar patterns, although the models designed specifically for live use invariably have a cardioid or hyper-cardioid response. Capacitor microphones should be considered as being slightly less robust than dynamic microphones (although not nearly as fragile as ribbon microphones), with less tolerance to dust, smoke and humidity, although dedicated live models are designed to be more impervious to such problems than studio mics.

Microphone Specifications

The sensitivity of a microphone is a measure of how much electrical signal you get out of it for a given level of sound input. This is usually expressed in millivolts-per-Pascal (mV/Pa), with one Pascal representing an air pressure of 10 microbars, equivalent to an SPL of 94dB. Studio capacitor mics may produce between 8 and 35mV/Pa—significantly higher than most dynamic mics—although it is not uncommon for live sound capacitor mics to have a deliberately curtailed sensitivity in order to more closely match dynamic models. Mic sensitivity is not really an issue when the sound sources are reasonably loud and close to the microphones, as is invariably the case in live performance.

The noise specification of a capacitor microphone is usually given as a 'self-noise' figure in dB SPL, which is the level of sound that would be needed to create an output signal at the same level as the mic's electrically generated background noise, or hiss. Obviously, the lower the self noise, the better—a good studio microphone might have self-noise between 5 and 15dB, but for live performance self-noise of up to 22dB is perfectly acceptable.

The maximum SPL that a capacitor microphone's active circuitry can accept before distortion is also important, especially where the sound source is very loud, such as a kick drum, guitar amplifier or close-miked brass instrument. A mic that can handle 135dB SPL can cope with just about any instrument, and for solely vocal use, a maximum SPL of 120 to 125dB should be able to cope with any voice.

The data sheets supplied with microphones can tell you a lot about their performance, once you know how to interpret them.

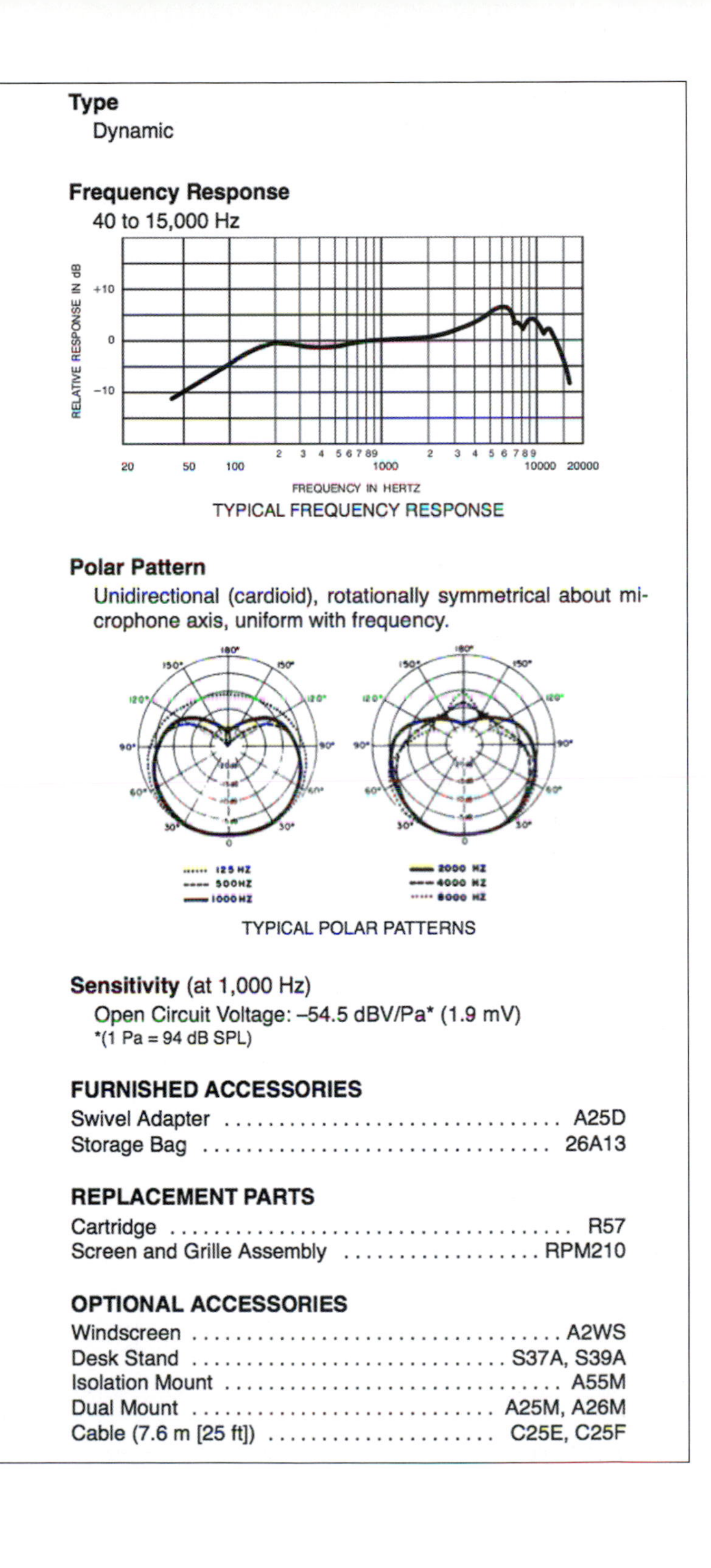

Type

Dynamic

Frequency Response

40 to 15,000 Hz

TYPICAL FREQUENCY RESPONSE

Polar Pattern

Unidirectional (cardioid), rotationally symmetrical about microphone axis, uniform with frequency.

TYPICAL POLAR PATTERNS

Sensitivity (at 1,000 Hz)

Open Circuit Voltage: –54.5 dBV/Pa* (1.9 mV)
*(1 Pa = 94 dB SPL)

FURNISHED ACCESSORIES

Swivel Adapter A25D
Storage Bag 26A13

REPLACEMENT PARTS

Cartridge R57
Screen and Grille Assembly RPM210

OPTIONAL ACCESSORIES

Windscreen A2WS
Desk Stand S37A, S39A
Isolation Mount A55M
Dual Mount A25M, A26M
Cable (7.6 m [25 ft]) C25E, C25F

Microphone Voicing

While a technically perfect microphone might have a perfectly flat frequency response, most microphones are 'voiced', with their response tailored to emphasise certain desirable characteristics in the sound. For example, vocal mics often have a built-in low-cut filter to reduce the amount of proximity effect, and most also have one or more 'presence peaks' in their frequency response in the upper midrange—the 3 to 6kHz region—which may be up to 6dB or so high, to help improve definition and intelligibility.

There was a time when Shure's SM58 was ubiquitous as the 'go-to' vocal mic for live performance, but there are now many worthy alternatives, not least in the form of Shure's own Beta 58 model. Because different vocal mics have subtly different presence characteristics, vocal mics are always best chosen to suit the voice of the performer, rather than on specifications alone. Female vocalists, in particular, often benefit from using capacitor or back-electret models because of their more natural high-frequency response.

Instrument mics may also have presence peaks, but usually have much less low frequency roll-off, as they may called upon to handle drums or bass instruments. They tend to have smaller

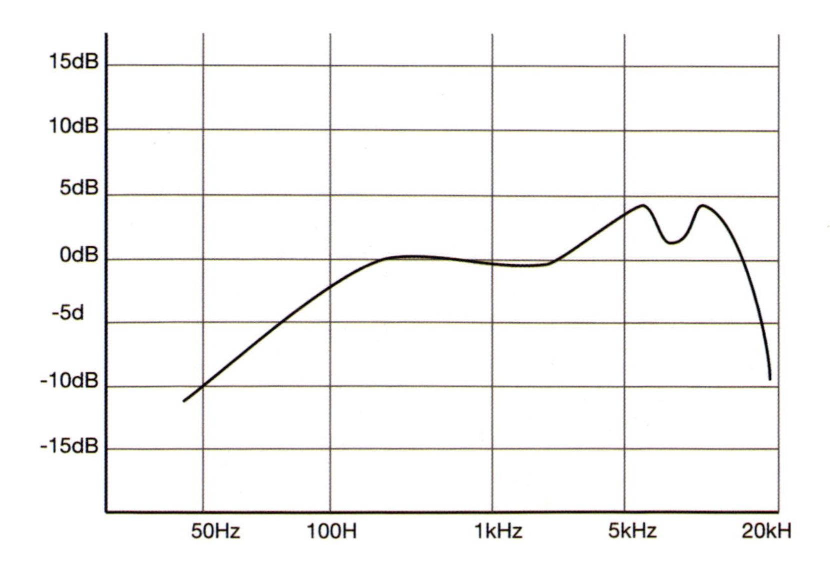

Vocal mics for stage use are often 'voiced' with presence peaks to help them cut through a live mix.

The classic, powerful vocal sound of Shure's long-lived SM58 design is still preferred by many vocalists, in spite of advances in the technology that offer greater fidelity.

head grilles than dedicated vocal mics as they don't need to resist vocal popping. The seemingly everlasting Shure SM57 is a good example of a general-purpose instrument mic often chosen for drums and guitar cabs, although there are many excellent alternatives from other manufacturers.

Kick drum mics are a special case, as you may find yourself needing a lot of EQ to get the right sound if you choose a mic with a flat response. This usually involves boosting the 80 to 100Hz region where most of the 'thump' of the drum resides,

Complete drum mic sets from a single manufacturer are now a popular choice.

boosting the upper mids where the click of the drum can be heard, and simultaneously cutting the lower mids to avoid muddiness or boxiness. Most dedicated kick drum mics, such as the popular AKG D112, have this kind of 'scooped' EQ curve built in. Furthermore, as kick drums are very loud, dedicated kick mics may be designed with lower sensitivity than a general-purpose dynamic microphone to handle higher source SPLs without overloading the mic preamp. A number of companies offer complete drum mic kits, often comprising a dedicated kick mic, three or four clip-on dynamic or back-electret mics for the snare and toms, and a pair of capacitor mics for the overheads. These kits vary quite widely in quality, but most can be persuaded to give acceptable results with a little EQ.

On-board mic switches

Personally, I hate stage mics that have on/off switches as it is possible for the singer to turn them off by accident. Some mics fitted with switches have a means of locking them in the on position so I'd advise doing this, and if you can't, tape over the switch.

Some microphones have switches for bringing in a 'Pad' and/or a low-cut filter—a Pad is an attenuator that reduces the mic's output by a fixed amount, usually 10 or 20dB, in order to avoid the mic preamp being overloaded by loud signals. You may also find a similar pad facility on the input of your mixer channel. If the signal level seems too loud, even with the channel gain turned down, switch on the mic's pad, if it it has one, or use the mixer's pad switch.

Some mics also have a low-cut switch to roll off the low-frequency response of the microphone and, again, there may often be a similar switch on your mixer's input channel. These switches are less common on live mics, but where they are fitted, it's usually best to turn the low-cut filter on for any sound source that isn't a kick drum or bass instruments as this helps exclude any extraneous low-frequency sounds that fall below the source's normal range, and also helps to avoid vocal popping.

Connections

Professional stage or studio microphones normally provide a balanced output on an XLR. Most professional mics are sold without the connecting cable, so these need to be purchased separately.

The robust and reliable 3-pin XLR is the *de facto* 'standard' connector for mics on stage.

Cheaper consumer microphones sometimes have an unbalanced output with a fixed cable terminating in an unbalanced TS jack plug. You can't plug an unbalanced mic into a balanced input that has phantom power applied to it as damage can occur, and in any event, consumer mics tend to be much higher in impedance than professional microphones. To summarise, use only 'proper' live sound mics with XLR outputs and avoid consumer microphones intended for Karaoke!

Chapter Six
DI Boxes

'DI' (Direct Injection) boxes are an essential and all too often underestimated component of any live sound rig, so it is worthwhile devoting a chapter to them.

Direct Injection simply means plugging an instrument's output directly into a mixing or recording system, via a suitable matching device. Any electronic instrument, such as a keyboard, drum machine, electronic drums and electric guitar/bass is a candidate for Direct Injection. The reason we need DI boxes rather than simply connecting the output of the instrument directly to the stage-box and multicore is that the

Passive and Active DI boxes from renowned Canadian manufacturer Radial Engineering.

output level, connection format and source impedance of the instrument may not always be compatible with driving long cable runs to a mixing console. The output also may not be tolerant of globally switched phantom power (although some acoustic instrument pickup systems and electric guitar preamps include active electronics capable of providing a balanced output without the need for a separate DI box, and some operate from phantom power).

Put simply, a DI box's purpose in life is to provide a convenient means of transferring the signal from an unbalanced, high-impedance instrument pickup, or from an electronically generated sound source such as a synthesiser, to a mixer's mic inputs as cleanly as possible in a balanced format and with a low enough impedance to be happy driving long multicore cables. The source signal level can vary enormously depending on the type of pickup or electronic source connected, so many models have switchable input sensitivity settings or pad switches. Since the instrument (or its amplifier) may well be grounded via a mains power supply, the possibility exists for a ground loop to form via the multicore and PA mixer's power supply, so most DI boxes also include a 'ground-lift' switch, to eliminate ground-loop hum in such situations. Most also include a 'Thru' socket that connects across the input socket, allowing the DI box to be used as a signal splitter, between the instrument and its amplifier, providing an isolated DI feed to the PA mixer at the same time as feeding the on-stage amplifier.

TIP: Note that the galvanic isolation of a transformer DI box doesn't necessarily provide a safeguard against electric shock during ground-fault conditions, as it is possible that the connected cabling is grounded at some other point at the source end of the system, or that the ground-lift is switched off. It is always wise to use RCD trips on backline amplifier mains supplies to ensure safety.

The piezo crystal-based pickups found on many acoustic instruments require a particularly high-input impedance (ideally two million ohms, or more) to work correctly, but most guitars fitted with this type of pickup also have matching preamps on board. Electric guitars and passive basses need to work at an impedance of between 250,000 and 1 million ohms—usually the higher the better—while synthesisers and other electronic sources are happy with anything above about 10,000 ohms.

As the stage multicore usually connects to the mixer's microphone inputs, it is usual for a DI box to produce a microphone-level output so that it can be plugged into the stage box alongside the normal microphone signals. DI'd signals don't suffer from spill in the same way that miked sources do, and there's less risk of acoustic feedback, although acoustic guitars, double basses, violins and so-on that have contact pickups may exhibit a degree of 'microphony' (microphone-like pickup due to sympathetic resonance) and so a measure of spill may still exist. Likewise, the resonant body of an acoustic stringed instrument such as an acoustic guitar can still lead to acoustic feedback problems as their vibrations couple back to the pickup system via the strings, although the feedback threshold will typically be much better than it would when using a microphone.

Although a DI is sometimes used with electric guitars, it is rarely taken directly from the instrument itself, as the amplifier is such a key element in creating the final sound. Bass guitars will often be DI'd directly, but many bass amplifiers are now fitted with a balanced DI output which allows them to be plugged directly into the stage box. However, some have only an unbalanced line output in which case a separate DI box is still needed. Guitar players who use a specialised preamp rather than a conventional combo-amplifier—such as one of the Line 6 POD family—may feed its output via a DI box to convert its line level outputs to balanced mic-level feeds that can be handled by the mixer's inputs.

Bass-guitar amplifiers often feature built-in DI outputs for sending a signal to the PA from the preamp stage rather than straight from the instrument.

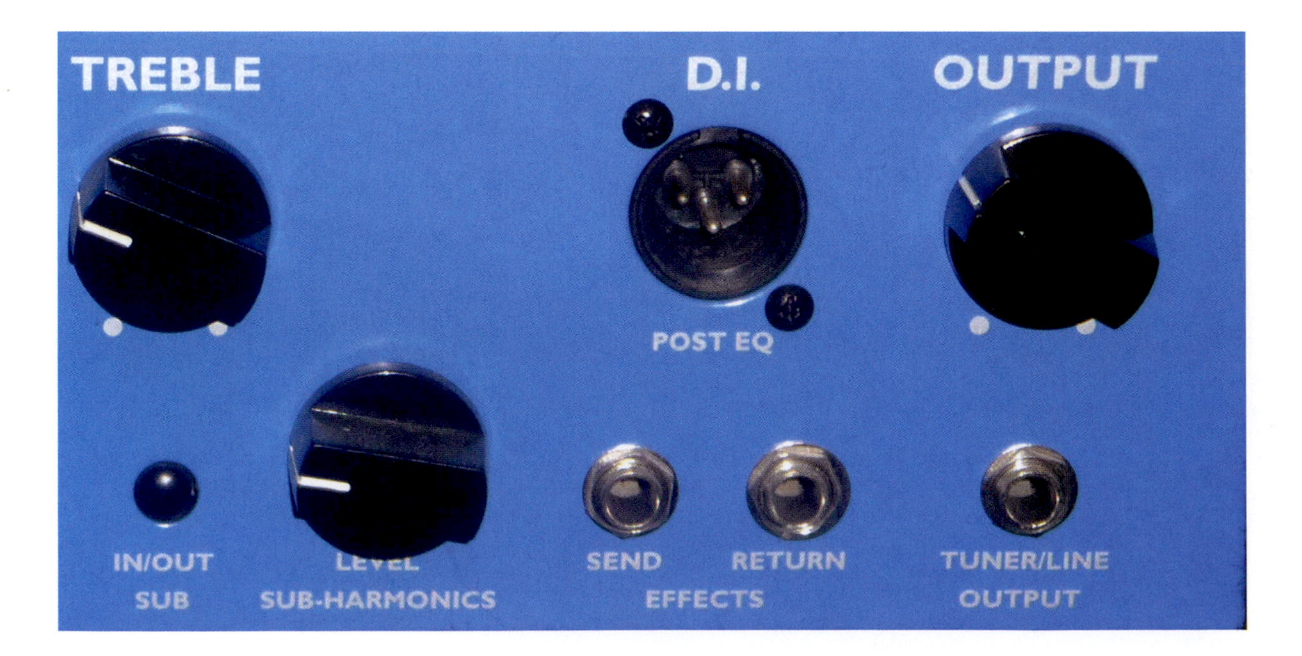

Most guitar players still prefer to use an amplifier on stage, and the usual strategy for reinforcement is to use a directional mic close to the speaker grille, as there's little spill penalty and the mic captures all the tonal nuances of the amplifier and speaker cabinet. There are, however, specialist DI devices that can accept the very high signal voltage from an amplifier's extension speaker outlet and turn that into a balanced, mic level DI feed, usually with some electronic filtering to emulate the frequency response of a typical guitar speaker cabinet. This filtering is known as 'speaker emulation'. Some guitar amplifier designers are now obliging enough to include a line-level output (often unbalanced) with speaker emulation built in, and you can connect this output to your mixer via a standard DI box.

In a similar way, keyboards and other electronic instruments may be DI'd directly, or a DI feed may be taken from the keyboard amplifier's line-out or DI socket. If taken from an unbalanced line-level output, the signal would normally go via a DI box to convert it to balanced mic level. As keyboard amplifiers are designed to reproduce a clean, uncoloured sound, there's usually no tonal penalty in DI'ing their line outputs directly into the PA.

DI Box Types

There are two main types of DI box: passive, transformer-based models and active electronic types (although most active electronic models also include a transformer). An audio transformer passes the signal magnetically between the primary

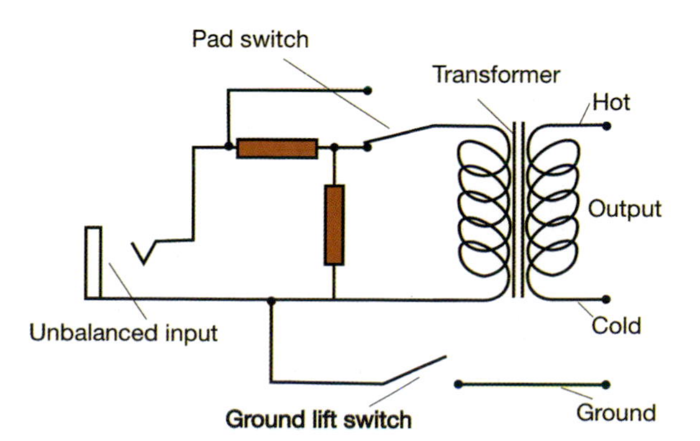

A transformer-based passive DI box.

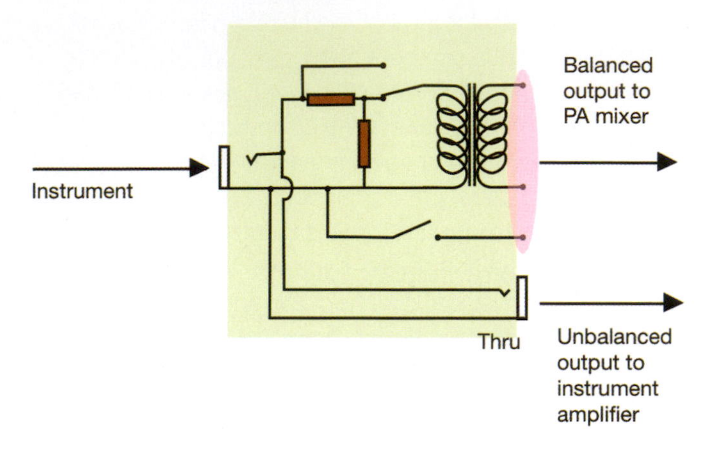

A DI box used to split a signal between a PA feed and an on-stage instrument amplifier.

(input) winding and the secondary (output) winding—there is no direct electrical connection between the two (called 'galvanic isolation'). In a passive DI box, the transformer has more turns of wire in the primary winding than in the secondary winding, which has the benefit of reducing the signal voltage from instrument to mic level, while simultaneously increasing the input impedance. With suitable wiring, the input accepts an unbalanced source, and the output is balanced—which is exactly what we need. However, the input impedance of a passive transformer DI box still isn't nearly as high as that of a typical active DI box, making passive, transformer-based DIs unsuitable for very high-impedance sources such as unbuffered piezo pickups.

Active DI Boxes

Active DI boxes may dispense with the transformers altogether, or combine the active circuitry and a transformer output stage to maintain galvanic isolation. The active circuitry allows the DI box to present a very high input impedance (often around 1 million ohms), affords more flexibility in matching input and output levels, and active filters can also be included. Transformerless designs still provide a balanced mic level output, but lack the electrical isolation of a transformer-equipped design.

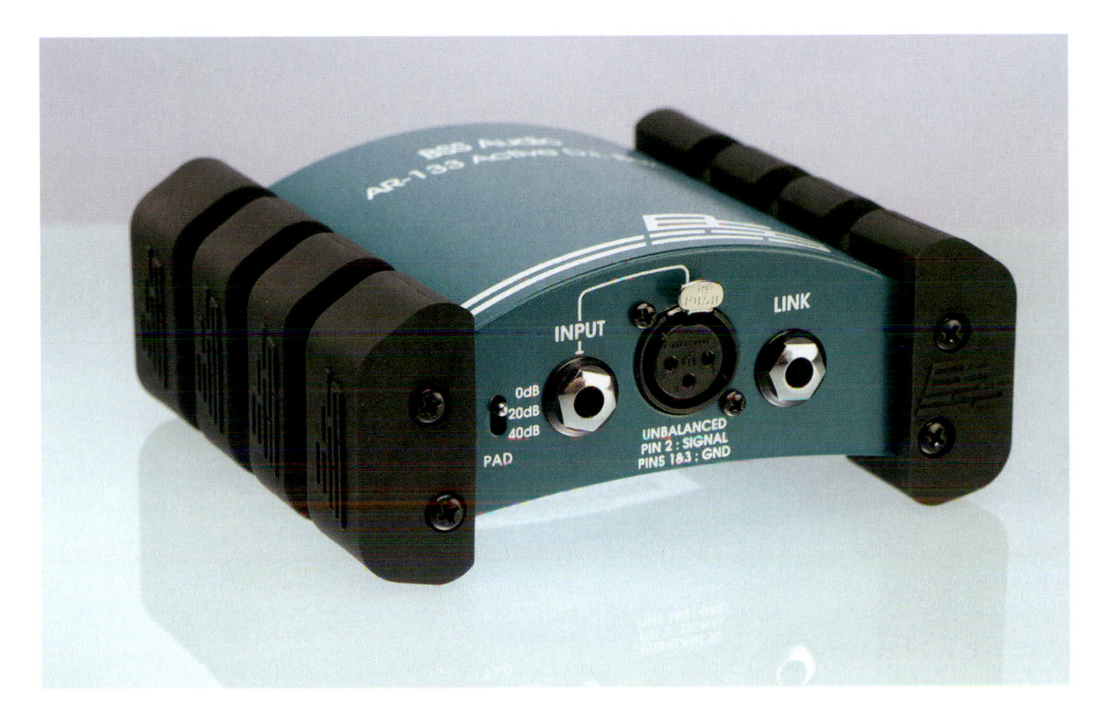

Active DI boxes require power from either a battery or phantom supply, but can usually handle a broader range of signals, with active buffering and filtering.

The active circuitry requires power and that can come either from internal batteries or phantom power. DI boxes that can use either usually switch off the internal battery automatically when phantom power is detected.

Speaker DI and Speaker Simulators

Sometimes it is beneficial to take a DI feed from the speaker output of an amplifier. There are both active and passive DI boxes designed to do this but, whichever you use, it must have an input connector specifically designated as being able to accept a speaker-level signal. Amplifier speaker outputs carry signals of several tens of volts, whilst line-level signals are usually only around a couple of volts, so plugging a speaker level signal into an active DI box's line level input would overload it and probably burn out its active circuitry.

Guitar speaker cabinets have a frequency response that is far from flat, usually with a number of resonances, and their limited audio bandwidth acts as a filter that is perfect for removing the raspy upper harmonics produced by overdriven amps or

This specialised DI box, optimised for guitar amps, can handle a speaker-level signal. It also features a switchable speaker-emulation facility.

distortion pedals, thus creating what we now accept as a typical electric guitar sound. The sound of a guitar is, therefore, quite significantly determined by the character of the loudspeaker cabinet through which it is played, so additional circuitry is necessary in the specialist DI box to emulate the distinctive frequency-response characteristics of a speaker cabinet.

Speaker emulators are sometimes built into guitar amps but more often they are to be found as an external device that combines the speaker-response filter with DI box features. It is very important with valve amplifiers to establish whether or not the speaker simulator also includes a 'dummy load' that matches the impedance of a guitar loudspeaker. If not, you must always connect a loudspeaker to the amplifier as well when it is feeding the DI box: failure to do so will leave the amplifier running into such a high impedance that it is effectively the same as having no load connected and,

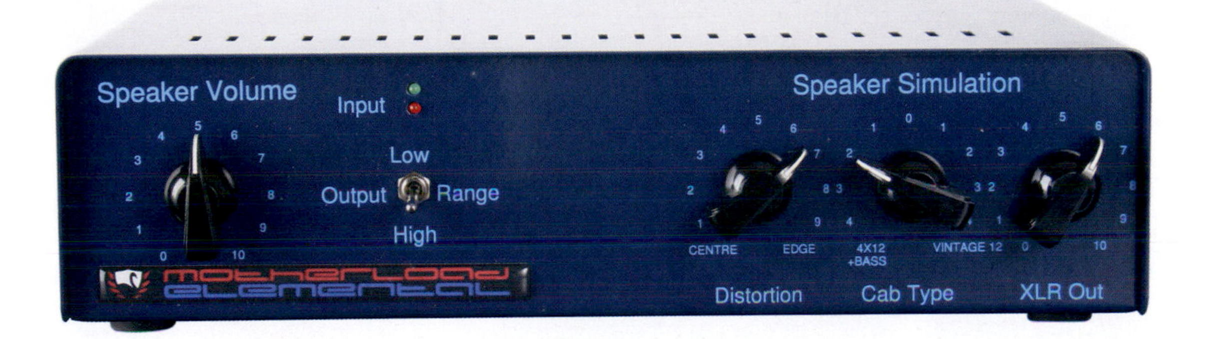

This speaker simulator emulates the frequency response of a typical guitar speaker and also correctly loads the output stage of the amplifier.

depending on the amplifier design, this could do serious damage to the output stage. Valve amps are not designed to operate with no speaker load connected. Solid-state power amps, on the other hand, are usually quite happy running with no speaker load.

Acoustic-instrument DI Considerations

DI'ing an acoustic instrument's pickup won't eliminate the risk of feedback altogether, but you'll certainly be able to get more level before feedback than when using a microphone. When feedback from, for example, a DI'd acoustic guitar does occur, it tends to be predominantly at the resonant frequency of the body, whereas when using mics the feedback frequencies depend more on interaction between the room, the foldback monitors, and the mic.

Using a third-octave graphic equaliser or a parametric equaliser to pull down the body resonance frequencies will allow more level before feedback becomes a problem, although the tone of the guitar may be affected to some degree. An automatic digital feedback suppressor that locates and notches out feedback frequencies can also help, and because the notches are very narrow, the effect on the overall tonality will be minimal.

Even in predominantly acoustic performances, acoustic instruments are often DI'd as well as or instead of miking, in order to achieve a more consistent signal into the PA.

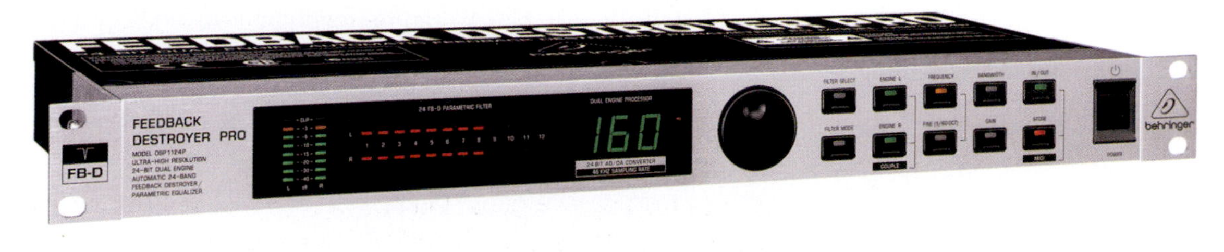

An automatic feedback eliminator will often have a less detrimental effect on an acoustic instrument signal than radical EQ to combat feedback.

Pickups And Tone

The sound of an acoustic instrument is a blend of the sound coming from the strings, from all parts of the body's surface, the sound-holes, and even from the neck, with reflections from nearby hard surfaces also contributing to the sound. Magnetic pickups that clip across the sound-hole are convenient for steel strung guitars, but they only pick up the string vibrations from a single location along the string and so tend to produce a sound that's more like a semi-acoustic electric guitar than a true acoustic (although some models also sense a little of the body vibration as well). Their output impedance is typically 5 to 10 kilohms, so an active DI box can handle them easily and, with a little EQ, they can give an acceptable sound in a band situation.

Piezo-electric transducers mounted under the bridge saddle or stuck inside the body tend to produce an inherently more 'acoustic' sound than a magnetic pickup, but they generally pick up vibrations from only one position on the body's surface, so the sound still won't be the same as the guitar heard acoustically. Piezo pickups, particularly when driven hard by loud playing, also have what some engineers refer to as a 'quacky' tonality that can never really be satisfactorily addressed via EQ or dynamics processing.

Piezo Pickup Impedance Matching

Piezo transducers are very high-impedance devices, so a dedicated preamplifier is often built into guitars, usually combined with EQ and volume controls, but plugging an unbuffered piezo pickup directly into an active DI box will give usable results in most cases.

Numerous manufacturers have designed dedicated acoustic guitar DI preamps that take the output from a guitar fitted with a piezo pickup, and apply EQ and other processing specifically designed to augment the sound of the acoustic guitar. Although often intended for unbuffered piezos, these can still benefit guitars with in-built preamps as they usually feature a variable frequency notch filter that can be used to attenuate the primary feedback frequency.

FISHMAN'S AURA DIGITAL PROCESSOR

A radically different and more sophisticated approach to creating a realistic acoustic guitar sound using a pickup is available in the form of the Aura products from Fishman—a company that has specialised in acoustic-instrument pickups for decades. Fishman's digital Aura process uses roughly 2,000 individual bands of EQ and phase adjustment to reconstruct the missing body resonances and complex tonal balance of a miked instrument. There's a range of responses available, as the Aura process works best if you use an 'acoustic image' that is very close to that of the actual instrument you are using.

Fishman's Aura process uses sophisticated digital EQ and phase manipulation to replicate the sound of a miked instrument from a pickup signal.

Most dedicated acoustic guitar amps usually include a DI output on a balanced XLR, allowing the sound of the guitar and any processing applied to it to be sent to the PA without the need for a separate DI box.

Ground-loop Hum

Ground-loops (sometimes called earth-loops) are created when an audio cable's screen links the grounds of two pieces of equipment—for example, between a guitar amp and a mixer—that are already independently grounded via their mains power cables.

This creates a closed loop, allowing ground currents to circulate, potentially resulting in audible hum or buzz. Disconnecting the mains safety ground from the mains plug of one of the devices may cure the hum, but creates a serious safety hazard and is never recommended. A far safer solution is to break the ground path via the audio cable, either by physically disconnecting the screen at one end of a balanced line-level connection or by using a line isolating transformer.

If a mains-powered instrument runs from a 'double insulated' external power supply, as many keyboards, guitar preamps and

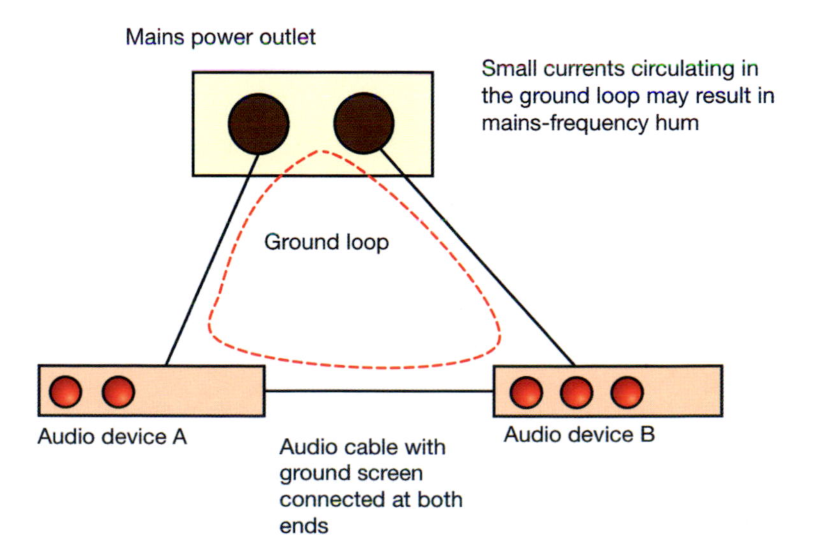

Having two paths to ground may cause a small current to flow in the ground connector, which causes mains-frequency hum in the output.

laptops do, or if the source is an instrument with a passive pickup, then it has no ground connection of its own and only becomes grounded via the device it is connected to. In such cases, the ground lift on the DI box should not be used, so that the input and output grounds are linked. This provides a proper ground connection from the mixer through to the source, which would otherwise be 'floating', without a ground reference, which can cause significant hum. If the ground-lift switch is opened on a DI box, there's no direct ground connection between the source and the mixing console, which is helpful if both are already grounded via their mains supplies as it will break the potential ground loop. It is essential that the source equipment is always correctly grounded.

TIP: Removing the ground conductor from the mains plug on one of the connected pieces of gear creates a potential safety hazard and is never recommended.

Less life threatening, but nevertheless important, is the fact that not all preamps and combos that claim to have a balanced DI output can tolerate phantom power being applied to their output sockets. Most of the reputable ones are fine, but if in doubt, always check the equipment handbook, and where that doesn't give you a positive answer, call the manufacturers. If it can't withstand phantom power then you'll still need to connect it via a DI box.

Computers

Laptop computers are increasingly being used in live performance to provide backing tracks, software instruments and even live effects, using either an audio interface or their integral audio outputs. For example, an audio interface with multiple outputs can be used to send a stereo track to the main PA and a click track or cue track into the monitors for the performers. Computer systems should always be thoroughly tested before relying on them at a gig, to ensure there's no nasty noises caused by the USB or Firewire connections breaking through into the connected audio interface. Running the computer from battery power during performance is the safest option in terms of noise—you just need to make sure the battery is properly charged.

Tablet computer systems—iPads, Android tablets etc.—must be thoroughly pre-tested to ensure there are no interference

If a computer plays any significant part in the running of a show, make sure the data is backed up, preferably more than just once. Hard drives can, and do, fail without warning.

problems. Conventional ground-loop problems are unlikely to affect laptops or tablets, even when running from mains power, as the PSUs are double-insulated devices that don't carry an electrical safety ground through to the device they are powering. Where noise does occur in poorly designed computers and audio interfaces it is due to interaction between the internal digital and analogue grounds.

If running on battery power doesn't cure the noise, you should try running the audio from the computer's headphone jack

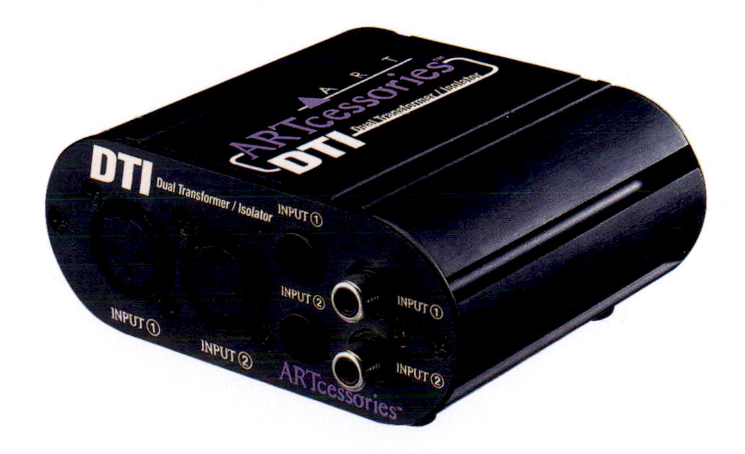

A transformer-based isolator can help eliminate the kind of unpredictable grounding issues that tend to show up when computers are used in live sound systems.

rather than from an external audio interface to see if that makes a difference. If it does you can then turn your attention to the interface and perhaps borrow a different model to see whether your interface is the cause of the problem or not.

If all else fails, noise problems can usually be cured by running a stereo audio feed from the computer's headphone output via a two-channel transformer isolation or DI box.

Chapter Seven
Other PA Components

The main PA components are, of course, the speakers, amplifiers, mixer, and microphones, but you'll need a few additional accessories to make the whole thing work, even if it's just a pair of speaker stands and a mic cable. This chapter looks at those extra bits and pieces.

At its simplest, a PA system comprises loudspeaker cabinets (usually a pair to provide a wider coverage than a single speaker), power amplifiers (either stand-alone, or built into the speakers or the mixer), and the mixer itself. Mixing is usually handled by a separate mixing console, although in the case of an acoustic guitar combo, a mic can often be plugged directly into the preamp section of the amp. A PA mixer need not be complex, but in addition to an adequate number of microphone inputs it should have basic tone controls and ideally provision for connecting external effects units and signal processors (unless these are already built in). It is also desirable to have some means of independently feeding a stage-monitoring system.

Mixers have a lot of vulnerable knobs and sliders so they need to be packaged securely in transit and the manufacturer's cardboard box isn't going to last forever. If you're not up to making your own box, then budget for a flightcase or a reinforced, well-padded mixer bag. Investing in speaker covers to stop your speakers getting beaten up in transit also makes sense, as tatty speakers are almost always the result of rough transportation, rather than actual use.

A very basic small PA system with monitoring.

Amplifier Loading

Issues such as impedance matching and power transfer are all taken care of for you when you buy active speakers, but if you prefer separate power amplifiers or powered mixers and passive speakers, you have to take on the responsibility of ensuring compatibility and optimum operation. To achieve the maximum power from an amplifier, it needs to be run into the lowest impedance load that it can safely handle, most often four ohms, although some very powerful professional models will drive loads as low as two ohms. When the speaker impedance is higher, the amount of current delivered will be lower and less power will be transferred. For example, an amplifier rated at 200 watts when feeding a four-ohm loudspeaker will only produce roughly half that if used with an eight-ohm speaker.

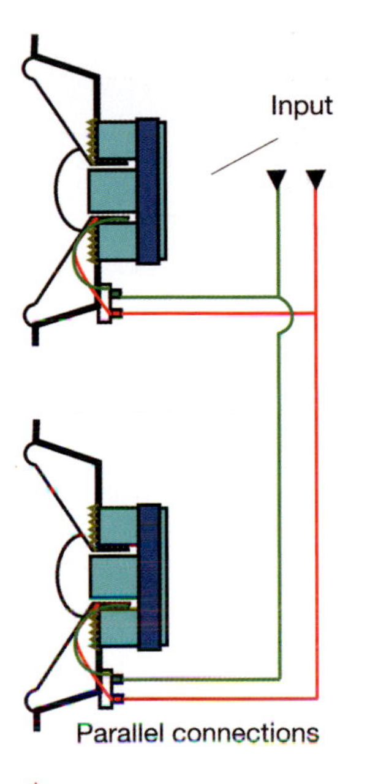

Two identical drivers connected in parallel will present a combined impedance equal to half that of a single speaker.

Where two identical speakers are fed in parallel from the same amplifier output, their combined impedance will be half that of an individual speaker: for example, two eight-ohm speakers wired in parallel present a combined impedance of four ohms. Some passive speaker cabinets have a parallel Thru connector allowing a link cable to be used to join the two speakers.

If two identical speakers are connected in series their impedance will double, but series wiring is not normally used to interconnect individual speaker cabinets. Cabinets containing two or more speakers may include internal series wiring or combinations of series and parallel wiring, but we don't generally need to be concerned about that: the overall impedance is all that matters, and that will be available in the manual or spec sheet, or even printed on the cabinet itself.

Cabling

For very small PA systems, a combined mixer and amplifier can be very convenient, the main disadvantage being that long leads are often required to feed the loudspeakers. Speaker cables for these systems need to be made from heavy-gauge wire as their resistance needs to be very low compared with the impedance of the speakers they are feeding. The higher their resistance, the more power will be wasted as heat in the cables and the lower the amplifier damping factor will be. Damping factor is essentially a measure of how tightly the amplifier can control the movement of the speakers to prevent overshoot due to the inertia of the cone. The higher the damping factor, the more tightly the speakers are controlled and the tighter the bottom-end response.

Guitar leads or other coaxial, screened cables should not be used as passive speaker cables in any loudspeaker system as their electrical resistance is too high.

Dedicated power amplifiers can usually be placed close to the speakers (in larger systems these are often to be found in a rack on or close to the stage), which helps keep the cables short, but if you're using a powered mixer and you also want to locate it in the optimum mixing position, part way back in the room, you'll need extra-long speaker cables, and the longer they are the heavier-gauge they need to be to maintain a low electrical resistance. For this reason, I'd normally only recommend powered mixers for self-mixing applications or for use in very small venues, where the mixer will be on or very close to the stage.

Speakon Connectors

Low-power, compact PA systems often use speaker cables terminating in quarter-inch jack plugs. but these should employ heavy-duty speaker cables and not instrument cables. It helps also to mark them as such, as I've seen musicians scratching their heads trying to locate a loud hum or buzz when they've mistakenly picked up an unscreened speaker lead and used it as an instrument cable.

However, jacks are not designed to handle high power levels, so higher powered systems invariably now use Speakon connectors. The Speakon connector was developed by Neutrik and is a locking connector designed specifically for passive speaker connections in which cables may be terminated either by soldering or via screw terminals, depending on the exact model of connector. The line connectors on the end of the cable mate with matching panel connectors, and the cable has identical connectors at both ends, so double-ended line couplers are used to join Speakon cables together if you need to extend a cable run.

Speakon connectors are inserted and then twisted clockwise to lock them in place, and vice-versa to release them. It has been known for those unfamiliar with Speakons to simply push them in and then wonder why they don't work—the contacts don't mate until the plug is twisted. Speakon connectors are available in two, four and even eight-pole configurations. An eight-pole connector is bigger than the more common two- and four-pole versions and is unlikely to be encountered in the type of systems discussed in this book. Panel connectors fitted with four-pole Speakons are commonly used, as they can carry separate woofer and high-frequency feeds from amplifiers driven via an active crossover, but they'll also work fine when only two connections are needed, such as when the speakers have their own internal passive crossovers. The connections of a four-pole Speakon are designated '+1', '–1', '+2' and '–2', with a two-wire connection using only the '+1' and '–1' contacts. Logically, a four-pole Speakon cable connector will not mate with a two-pole panel connector, which is why many manufacturers routinely fit their speakers with four-pole connectors, even if only two are used. That way either a two-or four-pole cable can be used.

Speakon connectors were developed specifically for higher-powered loudspeaker applications.

POWERCON CONNECTORS

Neutrik also developed a high-current mains power connector called a Powercon for use on equipment where the familiar IEC connector (the one many musicians still insist on calling a kettle lead!) isn't really up to the job. Powercons are locking connectors that can accept much heavier-duty mains cable than a standard IEC connector. Powercon plugs look superficially similar to Speakon plugs, but they cannot be interconnected by accident. While Speakons can be purchased at most music stores that sell PA equipment, I've usually been met with a blank look when asking for a Powercon, so if you have equipment that requires them, be sure to carry at least one spare, as if you find yourself at a gig with no Powercon cable, there's no way to get yourself out of trouble.

Mains Power Distribution

Mains power distribution boards are a must, as it is very rare for a typical pub or coffee shop to have enough power sockets to accommodate all the PA and backline gear, and even if they do, the sockets will almost certainly be in the wrong place. It is a good idea to always carry at least one long extension reel, in addition to your usual power strips, but it is very important that cable reels are fully unwound when you are drawing a lot of current through them. I've seen these reels melt and catch fire on more than one occasion when the PA or lighting operator (and it wasn't me, before you ask!) ignored the warning and left it coiled during operation. Most reels have their wound and unwound current rating printed on the side.

Modern portable sound equipment, and especially LED lighting, doesn't actually take a lot of current, but you should always do a rough calculation based on the actual power consumption of your equipment to help you estimate how many devices can be safely powered from a single wall socket. The PA as a whole will probably take the most power, so this should ideally be run from its own wall socket if the system is rated anywhere above a thousand watts. However, ensure that the backline equipment is fed from the same ring-main, and ideally from an adjacent or nearby socket, as this will reduce the risk of ground loops.

RCD Safety Trips

Not all venues can be guaranteed to be wired to the highest standard, and even where they are, a serious electrical fault might still create the risk of an electric shock. It's well worth checking the mains sockets you want to use at a venue with a simple plug-in socket tester before plugging in any of your own equipment. I've lost count of the number of times I've found sockets with missing earths or L-N reverses!

Plugging in each of your distribution boards via an RCD (Residual Current Device) is also strongly recommended to provide an extra level of safety. RCDs monitor the current in both the live and neutral wires, as this should be exactly equal if there's no leakage to ground anywhere in the system. If a fault

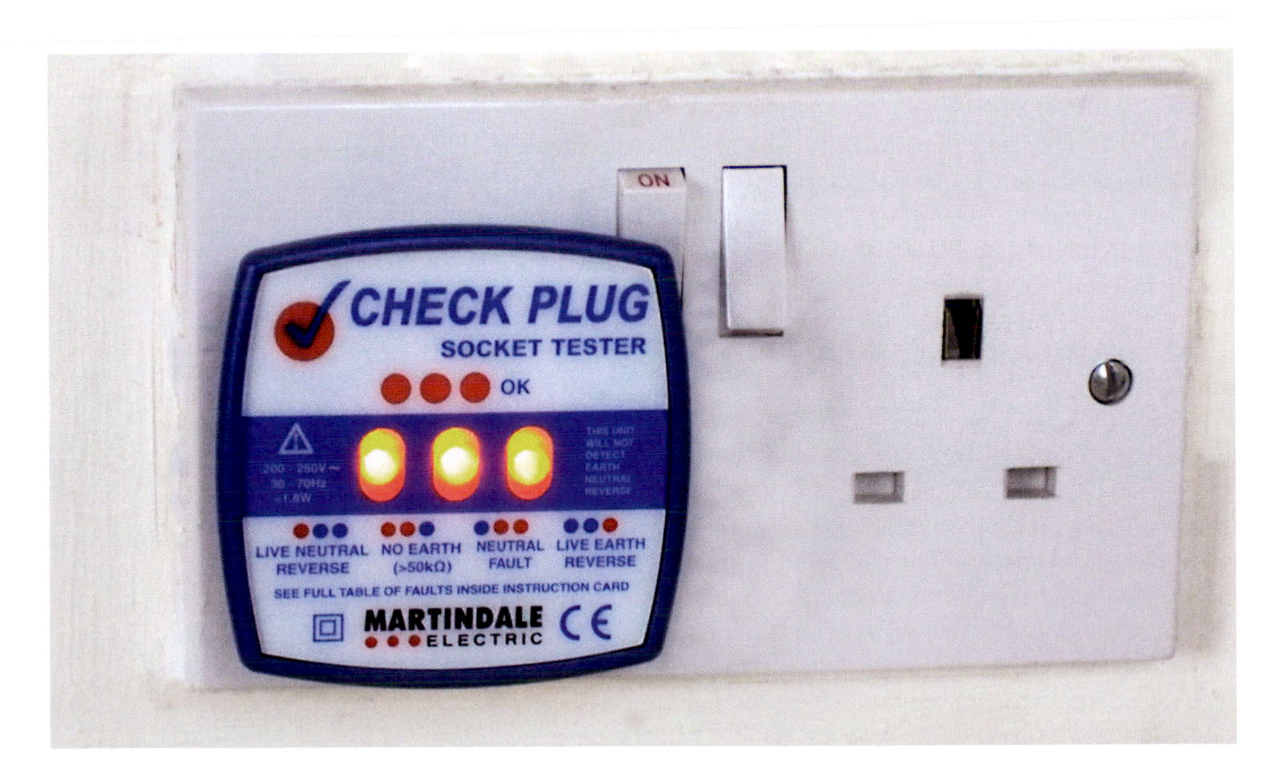

Mains socket testers are readily available and not expensive—it is well worth the effort of checking that the venue's power outlets are performing correctly before connecting your system to them.

Connecting each of your mains distribution boards via a Residual Current Device (RCD) should protect you from a potentially lethal electric shock in the event of a serious fault arising somewhere in the system.

arises that allows current to flow to ground—perhaps via the guitar player and his amplifier—the RCD will detect a difference between the live and neutral currents and disconnect the power before any serious harm is done. Getting your equipment and power cables safety-tested (PAT certified, in the UK) on a regular basis significantly reduces the risk of accidents, too.

Speaker Stands

Pole-mounted speakers are the most commonly used type for small-venue applications. The vast majority of these feature 35mm pole-mount sockets (sometimes called 'top hats') built into their bases. These are compatible with the telescopic poles that either have the familiar tripod bases or are mounted into the tops of subwoofers. My personal preference is for good quality, lightweight aluminium tripods, but, just as with mic stands, it is

Good-quality, stable stands are essential for correctly positioning compact PA cabinets so that they project properly.

Stands that can be cranked up after the speaker is mounted saves you doing the heavy lifting work.

better to spend a little more on models with metal fittings in the stress-bearing areas rather than plastic. Steel stands may be a little cheaper, but they are noticeably heavier. Almost all designs use a pin-and-hole system to provide a greater level of security once the correct pole height as been set.

Where the speakers are very heavy, as some 15-inch models can be, stands are available that can be cranked to height once the speakers have been lifted onto them at a lower safe-lifting height. They cost a little more but help avoid lifting accidents and back strain, so they may be a good investment. Always check the maximum load-bearing spec of your speaker stands to ensure that they can safely carry the speakers you intend to put on them. It can also be a good idea to wrap some high-visibility, striped yellow and black 'hazard' tape around the tripod legs to help people avoid tripping over them. Make sure the legs are spread adequately on a solid base to ensure there's no risk of toppling, too.

Stage Box

Stage box mounted on the outside of a multicore cable drum.

TIP: While some users simply coil their multicore by hand and pack it in a large box, this risks putting twists in the cable, and with a large multicore, it can be surprisingly heavy work. A better option is to use a purpose-built cable drum, which allows you to to deploy just as much cable as is needed. A stage box can be fitted on the outside of the reel.

Mixing consoles are often located remotely from the stage, so that the mix engineer can hear the performance as the audience hears it, but the signals from on-stage microphones and DI boxes still have to be routed there somehow. In the analogue world, this is done using a multicore cable, sometimes called a 'snake'. This is terminated at the stage end with a connector box, called a 'stage box', housing XLR sockets, with either a multi-pin connector or a set of individual 'tails' to connect to the mixing desk at the other end. Additional 'return' connectors on the stagebox are quite common to allow the mixer's main and foldback outputs to be routed back to the stage. The cables within the multicore are individually-screened, balanced pairs, which makes multicores both thick and heavy, especially when you need a long one that may have 24 feeds from the stage and eight returns.

Digital Alternatives

Sound systems employing digital components are often able to use a 'digital snake' instead of a heavy analogue multicore. The signals are converted to the digital domain at the stagebox and then multiplexed, allowing them to be sent down a single length of high-frequency coax, optical fibre or a network cable. These are much easier to rig and transport than a traditional multicore, and may be a very cost-effective alternative depending on the number of channels required and the other system components. Compact digital mixers that can be remotely controlled via tablet devices (using Wi-Fi) offer an even easier alternative, as they allow the mixer itself to be left on the stage, where the mics and DI boxes can be plugged in directly. See Chapter 4 for more details.

Microphone Accessories

Although mics are often stored in the flimsy plastic pouches that they tend to be supplied with, a dedicated mic storage bag or case that can hold a number of mics is always a safer option. These usually take the form of woven box-shaped bags with zipper-closing lids and foam inserts to accept the individual mics. This system isolates the mics from shocks and also makes it easier to do a 'head count' at the end of the gig.

Look after your mics by transporting and storing them with proper protection.

Each of your mics and DI boxes will need its own XLR cable, ideally with a few spares, and of course the spare cables can be used to extend your usual XLR cables in situations where they need to run over a greater distance. Active speakers use the same type of XLR cable, so you can end up carrying quite a lot of them. My solution is to connect them all end-to-end and to wind them onto a garden hose reel. This is an inexpensive and lightweight solution, big enough to hold perhaps 20 or more five-metre mic cables. You can also test all the cables in one go by connecting each end of the whole string to a cable tester, although if there is a fault you'll still have to go through them all individually to find it! I use Velcro straps both to secure the outermost connector and to prevent the reel from turning in transit, which keeps the cables from unwinding.

XLRs can be connected end-to-end and stored and transported on a garden hose reel.

While I have never noticed any significant difference in performance between different brands of mic cables in live applications, it is worth avoiding at least the very cheapest ones as their XLR connectors tend to be of inferior quality. The security of the cable clamp at the rear of the connector is of particular importance, as this area takes a lot of stress. Cables with conductive plastic insulation are the most resistant to kinking, although cables with a woven copper screen offer better resistance to interference. However, for vocalists who like to hold the mic, it is worth trying a few brands of cable to see which has the lowest handling noise, as some cables can be very noisy when moved or trodden on.

Mic Stands And Clips

Mic stands vary enormously in quality and price: cheaper ones tend to have plastic fittings that will almost certainly break after a few gigs, and the swivel connector holding the boom arm is

Mic stands and stand adaptors are available fairly cheaply, but this is one area where you do tend to get what you've paid for, so it is worth buying the best you can afford.

also particularly vulnerable. Solid-base stands are available, but most people seem to opt for boom stands with three fold-out legs, as these are flexible enough for vocal use, for positioning mics low down for kick drums and guitar amplifiers, and for suspending the overhead mics above a drum kit. Note that, for maximum stability, the boom arm should always be aligned in the direction of one of the stand's legs and not over the space between them.

More costly stands have sturdier metal fittings and spare parts will be available in the event of breakages. You could get everyone in the band to buy their own mic stand, so each member can decide how much they want to spend on it, but be prepared for the odd failure at an inopportune moment!

Most microphones come with mic clips, with some also including thread adaptors—annoyingly, there are two 'standard' sizes for the thread on the top of a mic stand! Plastic clips frequently get broken or lost, while thread adaptors just seem to evaporate when you're not watching them! Spares are not expensive, however, so make sure you carry some. Nothing looks worse than trying to fix a mic to a stand with Gaffa tape because the clip has broken. Well, I suppose a drummer wearing 'budgie smugglers' might be worse, but let's not go there . . .

Toolbox

Gear that you are relying on can, and will, go wrong, so it is prudent to always carry a basic toolkit with you. In addition to screwdrivers, pliers, wire cutters and a few spare fuses, jacks and XLR connectors, it is useful to include a bright LED torch and a gas-powered soldering iron. These have built-in igniters and heat up very quickly, making them ideal for emergency cable fixes, faulty guitar wiring and the like—just remember to pack a small reel of flux-cored solder, and a can of the correct fuel to use with it. And whatever you do, don't forget your Gaffa tape!

A cable tester that can check all the common cable types (jack, XLR and Speakon) also saves a lot of messing around with a multimeter—although a basic multimeter is also a great asset for general battery voltage and resistance checking. Alternatively, include a dedicated battery tester for AA and PP3 batteries, if you use battery-powered pedals or radio systems.

Your toolbox should also include a mains socket tester to warn if there's a missing ground or if the live and neutral wires are reversed, and given the number of times I've personally encountered such problems in venue wiring, these simple three-

LED testers are well worth the very small investment involved. You can buy them at most DIY stores or electrical dealers. If you do come across a faulty power point, don't try to fix it, just notify the venue owner and run an extension cable from a socket that tests as OK.

More For The List

Other useful things to take along to venues include rubber floor mats that can be used to cover cables that have to be run through public areas or across the stage where people are likely to walk, and a stand to hold your mixer, in case there isn't a suitable table you can borrow. Cable ties or Velcro straps can also be useful in keeping cables out of harm's way, and I have been known to carry screw-in hooks that can be fixed to the upper edge of a doorframe when the venue manager isn't looking to provide a secure way of running cables over the doorway.

Also, don't forget my comments on planning for gear failures—a guitar preamp such as a Pocket Pod can get you through the night if a guitar amp fails, while being able to redeploy a powered monitor or two to fill in for a dead PA can equally be a real gig-saver. The further your gig is from home, the more you need to be prepared for such eventualities.

If you use a computer on stage, invest in a laptop stand as this will be far more secure than trying to improvise using something not designed for the job. An MP3 player, or even a smartphone, for interlude music can also be useful, but make sure you have the appropriate cable for it. A folding sack-truck can make it much easier to transport heavy items in and out of venues, and of course you need a carpet for the drum kit to save the drummer following the kit around the stage.

Many small venues have no special performance lighting at all, and playing beneath a fluorescent strip light or energy-saving bulb tends to undermine the atmosphere you're trying to create! Colour-changing LED spotlights are getting less costly all the time, and have the advantages of being very compact and producing very little heat. The more portable systems comprise

Lightweight LED lamps have transformed the portable stage lighting world in recent years.

four flat lighting modules mounted on a lighting bar and stand, all of which folds up to around the size of a guitar case. These can be set to cycle automatically around the colours, to operate via a sound-to-light mode, or they may be fully remotely controlled via a proper DMX lighting controller.

Chapter Eight
Equalisers And Effects

In the recording studio, equalisers come in many different forms, but in smallish live-sound setups the choice is usually limited to the equaliser you have on your desk channels, and maybe a graphic equaliser on the main outputs. Graphic equalisers are very commonly used in live sound as they are easy to set up and can apply a more extensive degree of equalisation than the usual desk EQ.

Equaliser circuits involve some active circuitry to enable them to selectively boost or cut parts of the audio spectrum relative to other parts, so you can think of EQ as being a kind of frequency-selective volume control, or to put it in simple consumer-audio terms, a type of 'tone' control. High- and low-pass filters, on the other hand, can only reduce the level above or below their cut-off frequencies, and can be active or passive.

In a simple, first-order filter, the signal level drops by 6dB for every octave beyond the filter's cut-off frequency (an octave is simply a musical term for the doubling, or halving of frequency). So, for example, a signal two octaves down from the cut-off frequency of a 'high-pass' (also called a low-cut) filter will be attenuated by 12dB. Signals three octaves below will be attenuated by 18dB, and so on—the attenuation slope never levels out. Cascading (placing one after the other) two filter stages produces a second-order filter, with a 12dB per octave slope, and three cascaded stages (third order) would produce an 18dB/octave slope. These second and third order filters are very important in live sound and are included in many mixers in

Simple high- and low-cut filter response.

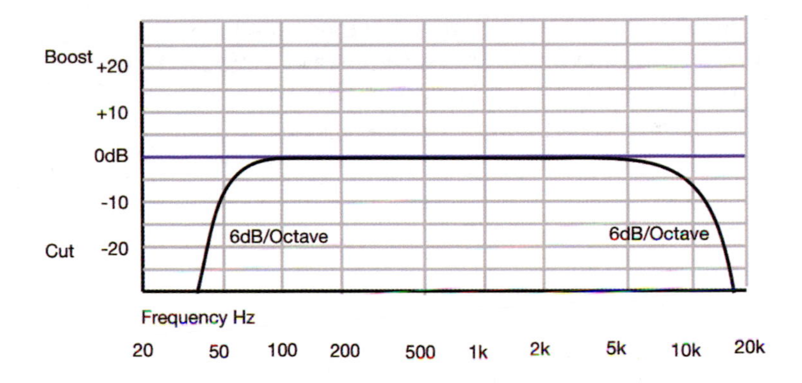

the form of a low-cut control, operating at around 80Hz, to get rid of unwanted low frequencies, such as popping and stand-borne stage vibrations, on vocal mic channels.

Shelving EQ

The high-and low-EQ controls on your mixer are most likely to be 'shelving' filters, although some mixers may offer fixed or switchable 'bell-curve' options too, which will be discussed shortly. Unlike the basic high-and low-pass filter, the shelving equaliser is based around active circuitry first developed by British engineer Peter Baxandall, the main characteristic of which is that the equalisation slope flattens out to a plateau beyond the cut-off frequency. All frequencies in the area where the shelf flattens out are therefore cut or boosted by the same amount. This type of equaliser has separate high and low gain

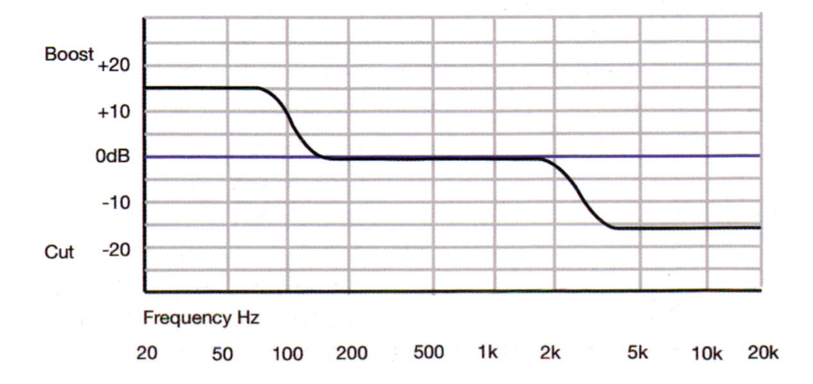

Shelving EQ can both cut and boost, with the slope flattening out beyond the cut-off frequency.

controls, each providing both cut and boost options, with the control having no effect in its centre or 'flat' position. A typical cut/boost range is between 12 and 18dB, depending on the circuit design. The diagram shows the low shelving EQ to be set to boost and the high shelving EQ set to cut.

Parametric Filters

While very few compact mixers actually feature a true parametric channel EQ, it is useful to know what the full-featured design has to offer, so that you can appreciate the compromises made in simplifying it. The conventional parametric EQ is a bandpass or 'bell-curve' (because the frequency response resembles the shape of a bell) equaliser centred around a specific frequency, with the boost or cut diminishing on either side of the centre frequency. A fully parametric EQ allows both the centre frequency and the width of the bell to be adjusted. We call the latter parameter 'bandwidth', defined as the width of the curve in Hz at the two points either side of the bell curve where the level is 3dB before the peak level. This type of equaliser can be designed to produce both cut and boost. The illustration shows one parametric band set to boost and another to cut.

You may also come across the term 'Q' in the context of equalisers where Q actually stands for Quality: it's not describing how well or otherwise the equaliser has been designed, but rather how 'tight' or narrow the filter response is!

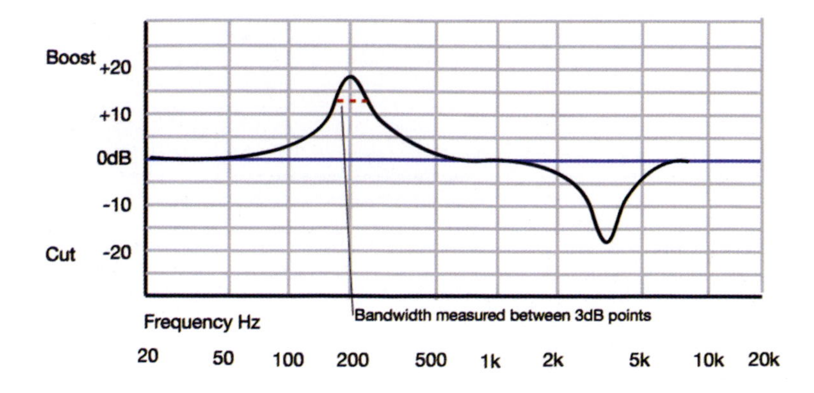

Parametric EQ, with variable gain, centre frequency and bandwidth.

Q is defined as the centre frequency of the filter divided by its bandwidth at the -3dB points, so the narrower the bandwidth, the higher the Q number. Throwing some numbers at you, a very broad bell-curve might span three octaves, which is a Q of 0.4, while a very sharp bell-curve might span just a quarter-octave, which is a Q of 5. A one-octave bandwidth is a Q of 1.4. A fully parametric equaliser typically provides between 12 to 18dB of cut or boost range, with three controls: gain, frequency and bandwidth.

Console EQ

Most small and medium-sized analogue mixing desks usually offer fixed-frequency high and low shelving controls, along with one or two frequency-adjustable bell-curve bands to cover the midrange. An additional fixed-frequency low-cut filter, usually set at around 70 to 80Hz, with a 12 or 18dB/octave slope, is often also included. The Q control of the fully parametric EQ is sometimes replaced with a switch offering a choice of two preset bandwidths, designated something like 'wide' and 'narrow'. Further down-market, the Q is fixed, with no switchable options, so the mid EQ band(s) have only two controls: cut/boost and frequency. This type of equaliser is

Typical compact mixer EQ with 'sweep' mid band.

often called a 'sweep EQ', allowing just the filter frequency and gain to be adjusted.

On even more compact designs, the midrange frequencies and bandwidths are both fixed, so all you get is a cut/boost control(s) at a fixed frequency. Where only a single midrange control is fitted, the frequency tends to be set at around 2.5kHz as that is a useful area for achieving vocal intelligibility. The smallest mixers of all, may even do away with the mid control altogether, so you are left with the high-and low-shelving controls only.

Graphic Equalisers

In live-sound applications, graphic equalisers are often used to help combat both feedback and room problems, such as modal resonances that occur at specific frequencies. Every room has resonances that are determined by the room's dimensions and the materials from which it is constructed, so to create a subjectively even tonal balance, you may need to use an equaliser to attenuate frequencies that correspond to the room resonances. Graphic equalisers are very intuitive to use for most people, because their multiple sliders create something of a visual representation of the overall tonal shape (although this is rarely entirely accurate!).

Behind the sliders of the control panel, a graphic equaliser is essentially a bank of narrow band-pass or bell equalisers, typically spaced at half- or third-octave intervals, although smaller graphic equalisers may have bands that are a whole octave wide. Logically, the more bands a graphic equaliser has, the finer the degree of control it offers. The individual filters are designed to overlap to provide a nominally flat response when they're all adjusted by the same amount. The two faders at the upper and lower extremes usually operate a shelving low and high equaliser, so you can think of the graphic EQ as being like a console EQ with a lot of mid-type controls, each one preset to a different frequency across the whole audio spectrum.

Graphic EQ sliders usually have a tactile centre notch, and moving upwards from centre gives boost while moving down

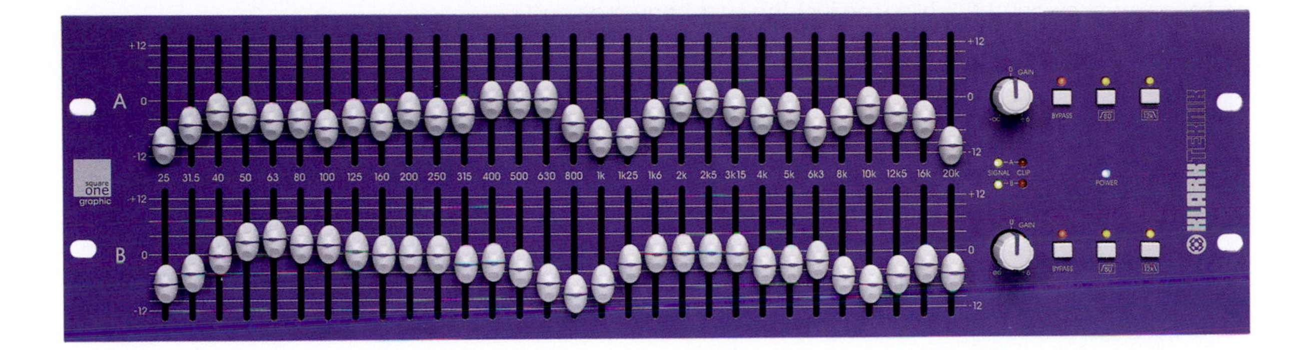

High-resolution, third-octave graphic EQ.

gives cut. In some graphic EQ designs the Q value for each band changes depending on how much cut or boost is applied, providing an asymmetrical filter characteristic that is narrower in cut mode than in boost mode. This approach takes into account that the human hearing system can readily tolerate very narrow cuts or notches created using EQ, but perceives narrow boosts as being unnatural-sounding.

Some models include peak LEDs above each fader that light up if feedback occurs in that band. These can be very helpful when 'ringing out' a system, as you get a visual indication of which fader needs pulling down to help cure the problem. Even very small PA mixers now sometimes include graphic equalisers, but these often have too few bands to make them effective in notching out troublesome feedback frequencies, although they can sometimes be useful for more general room EQ tasks.

Digital Mixer EQ

Unlike analogue designs, digital mixers don't have to provide panel space for all the EQ controls on every channel—many adopt the 'assignable' or 'fat channel' approach, with a single set of EQ controls used to adjust whichever channel is currently selected for attention. This means that digital mixers are able to offer far more sophisticated facilities than their analogue counterparts for a given physical size and, for example, even the low-cost Mackie DL1604 digital mixer, which is controlled entirely via an iPad, has fully parametric mid controls, as well as third-octave graphic equalisers on both the main and monitor outputs.

All the common analogue types of EQ have their digital counterparts, although digital processing can be designed to cover a much wider range of frequencies, as it isn't limited by physical filter components. Some even include sophisticated digital modelling processes to replicate 'imperfections' in the original analogue designs that may have contributed to their perceived 'musicality'.

Whenever the frequency response is adjusted using analogue components, such as capacitors and inductors, the phase shifts progressively for different frequencies—in effect, the harmonics that make up the signal are pushed out of alignment to some degree. While this sounds as though it might be a problem, the effect is actually very natural, we are are very accustomed to it, and it is often the reason why analogue equalisers are considered to sound 'nice' or characterful.

Conventional analogue equalisers are types of what we call 'minimum-phase' EQs. Minimum-phase behaviour can easily be replicated in digital equalisers, and for live-sound applications, it invariably is, although it is also possible to create a 'linear phase' digital equaliser, in which cut and boost can be applied without introducing different phase shifts at different frequencies. Linear-phase equalisers exhibit none of the characterful artefacts of conventional analogue EQ, but can sound rather 'clinical' in comparison, and are more appropriate to specific studio and mastering applications. Their complexity also comes at the price of an unavoidable processing delay,

SUBJECTIVE EQ PERFORMANCE

Different designs of analogue equaliser can sound quite different, depending on the approach taken by the circuit designer. The controls often interact to some degree and different EQ designs may also have different-shaped response curves, all of which contribute to their individual character. A whole market has evolved around reproducing classic outboard and console EQ circuitry, and you'll often hear people discussing the relative merits of different mixer EQs with users comparing the nuances of the 'American EQ' sound and the 'British EQ' sound.

again making them less suitable for live use, whereas in studio processing, any delays can be easily compensated.

Effects And Processors

Most PA systems usually include some means to add effects, with reverb and echo/delay being the essentials for pop music work. These effects can either come from an externally connected effects unit connected via the mixer's post-fade send and effects return channels or, as is very often the case now, from the mixer's own built-in effects processor. Not all systems include all the effects and processors described below, but a familiarity with the basics is always helpful, even if your mixer's effects section is based entirely on presets. At least you'll be in a better position to decide which preset to use.

Compressors

Compressors and limiters are often employed as a part of the live-sound audio chain in large-scale concert systems, but because they reduce the level difference between the loudest and quietest sounds, they can actually exacerbate feedback problems in smaller venues. A compressor is typically used to help keep an instrument or vocal performance at a more even level, although the 'side-effects' of compression can also be used as a creative effect—for example, to add punch to drums or bass, or to add sustain to a clean guitar sound. Whilst you can obviously compensate for level variations during performance to some extent by adjusting the mixer faders, a compressor circuit can always react more quickly than even the fastest human engineer's manual 'fader riding'.

Compressors monitor either the signal level by means of a circuit called the 'side-chain'. In a conventional compressor, the side-chain signal is constantly compared with a fixed threshold level set by the user, so that when the signal exceeds the threshold, the side-chain sends a control signal telling a 'gain cell', to turn the signal level down. A variation on this theme is to have a fixed internal threshold with the amount of compression determined by adjusting the input signal level instead.

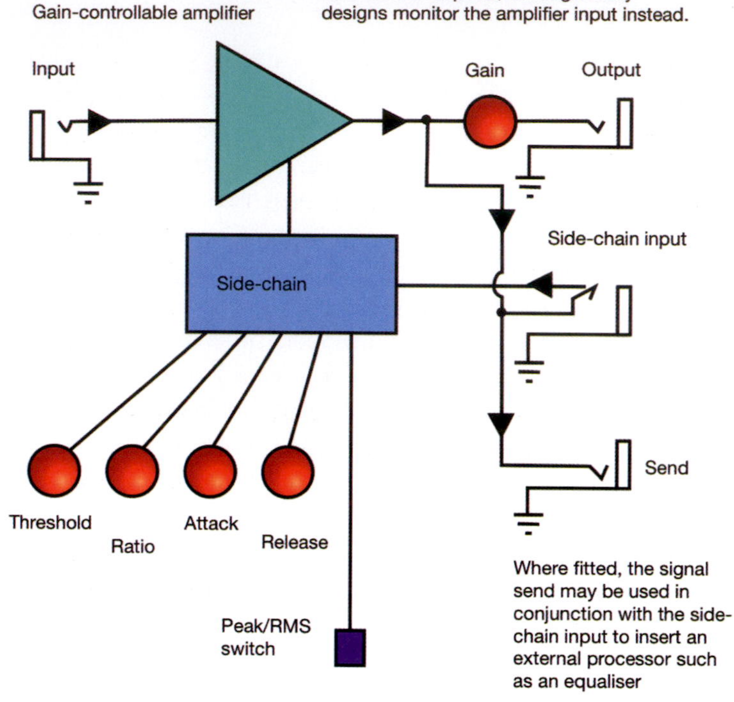

Compressor schematic diagram.

All the controls discussed below relate to a typical hardware compressor, although a compressor built into a mixer may have rather fewer controls: for example, many of Yamaha's smaller mixers have a compressor controlled by a single knob providing 'more' or 'less' compression, with all other parameters being automatically adjusted to optimal settings.

The level above which the compressor starts to apply gain reduction is known as the threshold, and the amount of gain reduction applied when the threshold is exceeded is set using the ratio control. The ratio simply tells you by how many decibels the input level has to rise (once the signal has exceeded the threshold level) in order to result in a 1dB rise in the output. For example, a 5:1 ratio means that if the input level rises 5dB above the threshold, the output increases by just 1dB—and if the input is 10dB above the threshold the output will increase by just 2dB. Realistic ratios used for gain control are between 2:1 and 5:1. Some designs have no ratio control at all, with the ratio simply continuing to increase the more the signal level exceeds the threshold.

In the simplest compressors, the sidechain is designed to respond to the average signal level, in much the same way that the human ear does. This is often referred to as an RMS (root mean square—a specifc way of deriving an average value) characteristic. However, when dealing with very short transients such as drum hits, a faster response is often desirable, so the compressor may have a switch to select Peak mode, which allows the side-chain to react to peak signal levels rather than to average levels.

Where audibly subtle compression is required, a 'soft-knee' compressor characteristic is often used. In a soft-knee compressor, instead of all the gain reduction kicking in as soon as the signal exceeds the threshold level, the ratio increases gradually as the signal level approaches the threshold so that the onset of gain reduction is far more progressive. A soft-knee compressor's ratio control usually sets the maximum ratio that will be achieved once the signal has passed the threshold.

The amount of gain reduction applied is usually shown on a meter, and where a two-channel compressor is to be used on a stereo mix or subgroup, it's important that the unit has a stereo-link facility. This prevents the loudest channel receiving more compression than the other channel, which would cause the stereo image to shift to the opposite side. When linked, a stereo compressor forces both channels to always apply the same amount of gain reduction, usually based on whichever channel's input is loudest at the time.

Attack And Release

The time it takes a compressor to react to signals exceeding the threshold level is separated into attack and release (or recovery) parameters, although attack would probably be less ambiguous if we called it 'response' or 'reaction time'. The attack setting determines how long the compressor takes to apply gain reduction once the signal exceeds the threshold, and release determines how long it takes the gain to return to normal again

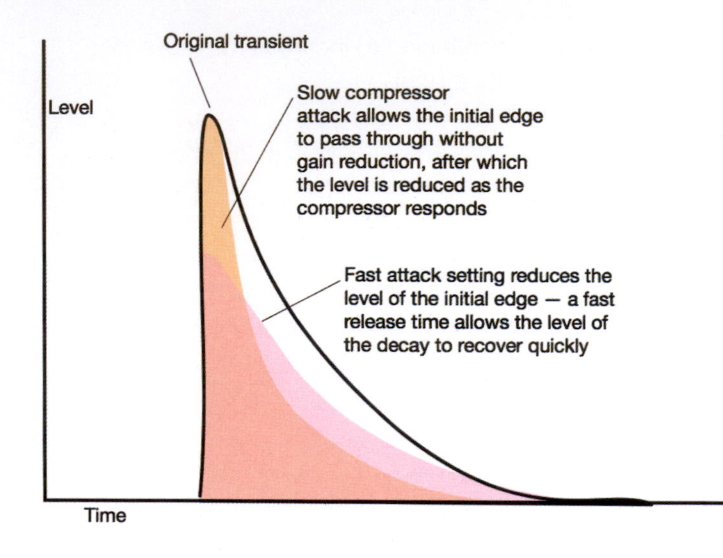

Percussive sound treated with fast and slow compressor attack times.

once the signal level has fallen back below the threshold. Compressors can be made to respond to level changes extremely quickly, but sometimes a slower response is what's needed. By setting a longer attack time, the start of a percussive sound, such as a drum hit or guitar note, can pass through the compressor unaltered before the gain reduction brings the main body of the sound under control, and this has the effect of enhancing the attack. A shorter attack time will bring the level under control more quickly. Release times are usually adjusted so that the compressor has time to reset to normal gain (ie. no gain reduction) before the next drum hit or note comes along. Very short release times are sometimes used to deliberately create noticeable 'gain pumping', giving an audible sense of energy as the compressor returns to normal gain over a short period of time.

Compressors with a lot of controls can be painstaking and time-consuming to set up, and even then the settings may not remain optimal if the dynamics of the input signal suddenly change. To get around this, some models of compressor have an auto-release option that allows the compressor to set a release time based on the dynamics and envelope of the audio signal. Other compressors may even be fully automatic, setting their own attack and release times based on the input signal's characteristics.

Make-up Gain

Compressors usually work by turning loud peaks down, but there's usually a 'make-up gain' control at the end of the chain and if you turn this up to attain the same peak level as before you applied compression, the quieter parts of the signal will now be louder. This results in a higher average level and the impression of a denser, more controlled sound.

Some compressors compensate for compression output-level reduction automatically, again a great time saver when setting up in a live environment, and single-knob compressors take care of pretty much everything, leaving you only to decide on whether you want 'more' or 'less' compression, albeit with the obvious compromise that you can't do anything about it if you don't agree with the predetermined settings.

Be careful with compressors in a live-performance environment. If you are adding 10dB of make-up gain to restore the same peak level you had before applying compression, quieter signals, below the threshold, are now being increased in level by 10dB, too. This will increase the level of any spill and, in small venues, this additional gain can be enough to keep your system permanently hovering on the verge of feedback, especially if you're compressing vocal mics. In practice, you should probably try to use as little compression as you can get away with.

Limiting

A limiter is essentially a fast-acting compressor with a very high ratio—20:1 or more—that prevents the signal from exceeding the threshold level. Limiters are often incorporated into crossovers and active speakers to prevent signal levels from getting high enough to cause clipping.

Side-chain Access

Some compressors allow external access to their side-chain, which enables the audio level to be controlled by the level of another audio source, as opposed to the compressor responding to the changing levels of its own input. A common example of external side-chaining can be heard every day on the radio when the music fades down automatically whenever the presenter speaks. This process is called 'ducking' and can be achieved by compressing the music to be ducked and feeding some of the presenter's mic signal, usually via a mixer aux send, into the compressor's side-chain input. Whenever the side-chain input exceeds the threshold setting, the music is reduced in volume. The amount of attenuation depends on the ratio control and the level of the presenter's voice signal.

It is also possible to insert an equaliser into the side-chain so that the compressor responds more assertively to whatever frequencies have been boosted (and ignores those that have been cut). A basic de-esser works this way: the compressor side-chain is fed with a signal in which the sibilant frequency range (typically around 6kHz) has been boosted. In this way the compressor ignores most of the input signal, but sibilant sounds will exceed the threshold and so the gain is reduced to help disguise the splashy 'ssss' sound.

Gates And Expanders

Noise gates, or more often just 'gates', are often employed in live performance, their purpose being to attenuate or mute signals that fall below a user-determined threshold level. Most gates have attack and release controls that allow the gating action to be optimised for different types of signal.

Fast attack settings cause the gate to open almost instantly when the input signal exceeds the threshold, allowing percussive transients to pass through unscathed, while a slower attack forces the gate to open more gradually. Very fast attack times, which are invariably used for percussion, can cause audible 'clicking' on non-percussive sources, so it is customary to slow the attack time to a few milliseconds when treating sounds with a slower onset, such as vocals.

A variable release time forces the gate to close gradually rather than snapping shut as soon as the signal falls below the

GATE SIDE-CHAIN FILTERS

Many noise gates now also include 'side-chain filtering' within their facilities. Using adjustable high- and low-pass filters, you can tune the gate response so that it opens only when it 'hears' something in the band of frequencies between the two filters. This makes it much easier to avoid 'false triggering' caused by loud spill signals occurring in a different part of the frequency spectrum from the one you are gating—for example, the kick drum false-triggering the snare drum gate. Don't worry too much if the gates are occasionally triggered by hits to other drums in the kit as this is inevitable given the close proximity of the drums. Note that the filters only affect how the side-chain responds—they have no direct effect on the audio signal passing through the gate.

threshold level, which helps to preserve the natural sound envelope of instruments that have a progressive decay, such as guitars and pianos. Faster release times can be used to deliberately shorten sounds, for example, to kill some of the ringing you might hear when close-miking drums.

Range, Or Depth

It isn't always desirable to have a gate mute the signal completely when it closes, as the sound of background spill switching on and off as well can sometimes be more distracting than having it remain audible. To get around this, some gates have an extra control that sets the amount of attenuation when the gate is closed. A reduction of around 10dB is often plenty to clean up a signal, without making the presence of a gate too obvious.

There are also devices called downward expanders that operate in a similar way to a gate except that the attenuation is set by a ratio control, much like a compressor in reverse, so that the gain reduction is more progressive when the signal falls below the threshold.

An EQ in the side-chain can help avoid false-triggering of gates and expanders.

Harmonic Enhancers

Equalisers cut or boost frequencies that are already present in the source material, but a harmonic enhancer generates new high frequencies related to the existing mid-range signals, making the sound clearer and brighter than it originally was. Different manufacturers use slightly different operating principles —Aphex's Aural Exciter, which pioneered this process, uses a combination of high-pass filtering, harmonic generation through distortion, and compression. The filter, which is usually adjustable, determines what part of the audio spectrum is used to feed the harmonic generator, and the new harmonics may be compressed before being mixed back in

TIP: Careful use of an enhancer can add life and clarity to dull signals, but over-use will quickly lead to a harsh and fatiguing sound. It's very easy to overdo the use of an enhancer without realising it, so frequent checking against the bypassed sound when setting up is always a good idea!

Harmonic exciters can create artifical, harmonically related high-frequencies to help boost dull signals.

with the original signal. As long as the level of added harmonics isn't too high, the human ear interprets them as being a natural part of the signal.

Connecting Processors

Gates, expander, limiters, compressors, equalisers, enhancers and pitch correctors all fall into the general category of 'processors' in so far as they process the whole of the signal, and are usually introduced via mixer insert points, rather than being added to the dry signal using a send/return loop. It is also possible to connect processors via a send-and-return loop if necessary, using a pre-fade send with the direct signal turned down using the channel fader. The 'no added dry signal' rule is often broken in the studio to create parallel-compression effects, but this isn't something I'd ever recommend for live use in smaller venues, as the heavy compression settings required could pose a serious feedback risk.

Effects For The Live Stage

Effects—often shortened to just FX —usually encompass things like reverb, delay/echo, chorus, flanging, rotary-speaker emulation, vibrato, pitch shift etc and so on, though reverb and delay are the most useful vocal treatments. Effects will generally be connected via the (post-fade) aux send/return connections of the mixer, or via the console's insert points. When fed from aux sends, the dry signal path through the effect should be turned off so that the output is 100 percent 'wet', but when used via insert points, the dry/wet effect balance must be adjusted on the effects unit itself.

Delay/Echo

Delay/echo for live performance is now almost exclusively taken care of using digital processors, and some units allow the delay time to be modulated to create effects such as chorus, vibrato, flanging and phasing.

The simplest delay effect is a single repeat using no modulation and no feedback. Short delays of between 70 and 120ms produce 'rockabilly-style' slap-back echoes, not unlike a coarse reverb, but longer delays produce a distinct single echo. Feeding some of the delayed output back to the input creates multiple delays that decay in level. Adding feedback to a multi-tap delay causes the echo decay pattern to become quite complex, again creating a pseudo-reverb effect.

While a digital delay can generate repeats that are indistinguishable from the original sound, having slightly 'dirtier' repeats can help them sit behind the dry sound in a more natural way. Emulations of tape and early analogue delays

Tape echo units may have given way to digital technology, but most of the digital units include an emulation of the imperfections of tape-based units.

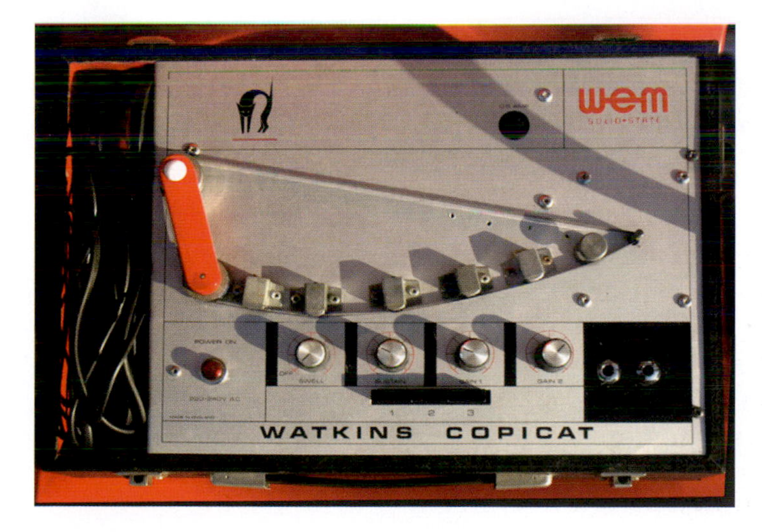

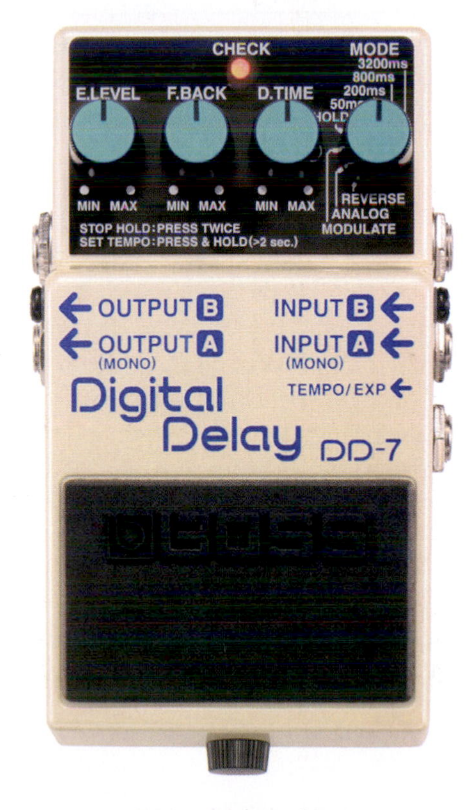

include filtering, distortion and sometimes pitch modulation to replicate the way these early devices sounded, and tape-delay emulation is still often chosen as the best-sounding effect for general purpose vocal and instrument use.

Delays are often timed to coincide with the tempo of the song, and a number of delay boxes and mixer effects sections include a 'tap tempo' facility: you simply tap a button at the tempo of the song and the delay time will be set to the time interval between hits. This is a really quick and easy way to make sure your repeats are in time with the song, so when choosing a mixer with inbuilt effects, check to see if tap tempo is available.

Delay Modulation

If you modulate the delay time the pitch of the delayed signal wavers alternately sharp and flat. The rate and depth of modulation can usually be adjusted, and this can be very subtle, as when replicating the wow and flutter of a worn tape machine—a beautiful effect when used subtly—or very obvious, such as when creating a deliberate vibrato. Vibrato can be created by using a very short delay and switching off the dry sound so you only hear the delayed sound. Usually the delay can be set to just a few milliseconds so it isn't perceived as a delay at all.

Phasing

Phasing is produced in a similar way to vibrato except now an equal mix of dry and delayed sound is used. As you adjust the modulation speed and depth, you'll hear the individual harmonics that make up the sound being emphasised in a dynamic and complex way. Feedback can be added to strengthen the effect, and there may be a 'polarity-invert' function to change the tonality and final harmonic structure, so try both positions to see what works for you. Phasing is a popular effect with both guitar and keyboard players, but will cause feedback problems if you try to use it on live vocals with the feedback control turned up.

Flanging

Flanging was originally produced in the studio by mixing together the outputs of two tape machines, each carrying the same track. Analogue tape machines never stay perfectly in time for long, especially if you give them a helping hand by dragging your fingers along one of the tape reel flanges, and as the timing between the machines begins to drift, you hear a 'phasing' effect, with deep comb-filtering caused by the small time delay between the two tape-machine signals. The familiar 'whooshing' sound was a popular effect back in the days of psychedelia where it was often applied to sections of a complete mix—very 'trippy'.

Tape flanging may be approximated digitally by adding a delayed signal to the original sound, then modulating the delay time of the delayed signal from near zero to several tens of milliseconds. Although the general settings for flanging are similar to those used for phasing, a longer delay time with a wider sweep range is normally used. Again, feedback can be added to make the effect more pronounced. As with phasing, inverting the polarity of the signal fed back to the input allows different harmonics to be accentuated by the filtering process. Try both options and see which you prefer.

Chorus

Chorus, again, uses a delayed sound with modulation in the same way as flanging and phasing, but usually with little or no feedback applied. Chorus creates the illusion of two instruments playing together by introducing slight timing and pitching differences between the dry and delayed sounds. By setting a delay time of between 30 and 70ms, the effect of the timing differences between the wet and dry sounds is pronounced enough, but still not long enough to produce an obvious echo. Many flangers have sufficient adjustment range to allow them to produce credible chorus effects, too.

Chorus was first developed for use on electric guitars and string synthesisers, although it is also commonly used by fretless-bass players and often added to synth pads. I've never found the need to add any of these modulation effects at the mixer in my PA work, as most players look after their own instrument effects, but it's useful to know about them just in case a situation arises where you need them.

Pitch-Shifting

Pitch-shifting changes the pitch of a sound without changing its duration, and while large amounts of pitch-shift that turn voices into demons or cartoon mice are of limited use in a musical context, subtle de-tunings of a few cents can make a very musical alternative to a chorus effect, adding texture to a vocal or instrument. The range of a typical pitch-shifter is usually one or two octaves up or down, with a fine-tuning range of a semitone. Some processors allow simultaneous shifting up and down in pitch,—a popular setting is +6 cents and –7 cents—to create a thickening effect that works well on guitars, keyboards and backing vocals. The main difference between this treatment and chorus is that it achieves its 'multiplier' effect with no delay modulation being audible.

Reverberation

Reverberation, or reverb for short, is a very familiar, natural effect heard in large buildings and big spaces with reflective surfaces. The resulting reflection patterns are immensely complex, but digital processors make short work of emulating the real thing. Most digital reverb units include a choice of halls, rooms, 'echo-chamber' and studio plate emulations, with some also replicating the sound of the spring reverbs used in guitar amps. There may also be non-natural reverbs, such as 'gated'—a short burst of reverb that ends abruptly—which can sound good on drums, especially if you remember the 1980s; and 'reverse', which is more of a special effect that goes some way towards replicating the sound of reverb building up before the sound that created it.

A basic reverb unit or pedal will include a means of selecting the reverb type, a reverb decay control, which sets how long the reverb hangs on for, and a wet/dry mix control.

More sophisticated studio models can include many more seemingly arcane adjustments that would require a couple of pages to explain (early reflections, diffusion, spin and wander), but reverb treatments used in live sound seldom need deep editing. You may come across controls for pre-delay (a short time delay between the onset of the reverb), and perhaps a parameter to affect the overall tonality of the reverb, but for PA purposes, the basic controls are often enough, especially when you consider that whatever reverb you add is also going to be coloured by the natural acoustic of the venue.

Pitch-Correction

The use of off-line and real-time pitch-correction has caused quite a controversy in the music business but, like the atomic bomb, you can't 'un-invent' it! Antares Audio Technologies developed the original Auto-Tune process, but there are now many products from companies such as TC Electronic and Digitech, for example, that work on a similar principle. This type of pitch-corrector works by first analysing and tracking the pitch of the audio input (which has to be monophonic and reasonably free from spill), and this 'tracked' pitch is then compared with the notes in a user-specified musical scale. Pitch-shifting is then applied as necessary to make the input signal match the pitch of the nearest scale note.

If the pitch-correction process is applied too efficiently, all human inflection is stripped from the performance and you end up with the familiar and much abused robotic, vocoder-like sound. In fact, some pitch-correction units have been produced to maximise this particular sound as a special effect! Used more subtly, however, automatic pitch-correction is capable of delivering reasonably natural-sounding results. A speed control determines how quickly off-pitch notes are brought into line, and by adjusting this so that the correction process is never applied too quickly, the natural slurs, inflections and vibrato are all retained while sustained notes are pulled into pitch.

Where the pitch-correction device is being controlled by the PA engineer, it is best to set up patches to suit each song with the patch including the appropriate scale notes (all the different

Personal pitch-correction pedals like this are designed to be controlled by vocalists themselves.

notes the singer needs to hit during the song), and other settings such as the correction speed. Personal pitch-correction boxes are also available that can be patched between the performer's mic and the mixer and engaged in the same way as a typical guitar pedal using a footswitch. Most of the the simpler ones are preset to a chromatic scale, with a simple 'more or less' knob to set the severity of the pitch-correction. They may also include basic vocal effects such as delay, reverb, EQ and compression. Some of the more elaborate ones also include automatic harmony generation.

While there's a lot you can do in the studio to optimise the use of pitch-correction, such as automating the speed control or bypassing the effect for all but the few notes that need it, live sound applications need to be handled in a simpler way. It almost goes without saying that the automatic approach works best when the vocal performance is already good, perhaps just needing a little push here and there to nudge it towards perfection. In fact, it can really only work at all effectively when the singer's pitching is reasonably good in the first place, for if the sung note is off pitch—perhaps because you are not hearing yourself very well—the correction process may actually force it to the wrong note altogether, making the performance sound even worse. Deliberate, slow pitch slides can also fall foul of pitch correction.

There's also the issue of hearing the pitch-shifted vocal sound coming back over the monitors while you are singing, which can cause a difference-frequency 'beating' effect in your head. Some singers use this beating effect to help guide their pitching, while others find it so off-putting they can't perform properly at all. My own experience is that if the amount of pitch-correction is moderate you can live with the beating effect, but if you set the correction to be too assertive, the result is really distracting. You also need to be aware of the fact that any spill picked up in the vocal mic will also be pitch-shifted along with the vocal, and if there's too much spill the device won't work reliably at all.

Auto-harmony

Auto-harmony pedals work on a similar technical principle to pitch-correctors, but in this application the pitch-shifting is used to move copies of the original notes to new pitches, according to the scale information in the current preset or user setting. This requires a degree of musical knowledge and patience in order to program different presets for different songs, as you have to specify the key of the song and the type of harmony used on the various parts.

Where you need something simpler and largely automatic, there are harmony devices that can read the pitch from a six-string guitar input (they have guitar in and thru jacks, so the guitar can still feed your amp). These are able to work out appropriate vocal harmonies in real time, based on the detected chords and can work surprisingly well.

All the units I've looked at myself actually track the vocal pitch really well and generate reliable harmonies, provided that the

Some of the more sophisticated vocal-harmony devices can analyse an audio input from a guitar and automatically generate the correct harmonies for the chords being played.

original vocal performance is reasonably good. The subjective quality of the harmonies seems to vary depending on the character of the singer's voice—female vocals seem to sound more natural than growly male vocals—but you really need to try one to see if it is going to work for you. Used with care, so the 'fake' harmonies sit behind the main vocal rather than fighting with it, the end result can be quite impressive.

Acoustic Feedback—Causes And Cures

When playing small venues, acoustic feedback can pose a serious problem, so it is well worth looking at its causes and at strategies for avoiding it, before looking at electronic remedies. In any venue, the microphones used in a PA system will pick up not only the sounds of the instruments or singers at which they are directed, but also other sound from around the room, including the sound of the PA system itself, either directly or reflected from the room's surfaces. As a rule, the closer the PA speakers are to the mics, the more signal finds its way back to them, and if too much of the sound from the PA speakers leaks back into the microphone, it will circulate around the system growing louder every time it goes around the loop, quickly building up into a continuous whine or whistle known as feedback or a 'howlround'.

It is very important to understand that the onset of feedback is linked to the overall system gain, not to the absolute level you hear from the PA speakers. That's why a very loud singer working close to a mic may have no feedback problems while a singer with a quieter voice, or one not working as close to the mic, in the same venue using the same PA gear, may experience significant problems. The quieter singer needs more gain to bring their voice to the same level, and as the gain increases, so too does the risk of feedback.

Nice pose, but that hand position on the mic will be wreaking havoc with its directionality, bringing feedback into play!

If the PA system appears to 'ring' when you speak into a microphone, it is a warning sign that full-on feedback isn't far away. The more of the sound from the speakers that is able to find its way back to the microphone(s), either directly or via reflections from walls and ceilings, the less gain you can use before acoustic feedback becomes an issue.

So what practical measures can be taken to keep feedback problems to a minimum? Well, firstly, get the vocalists to sing very close to their microphones. The closer your source, the higher the output, so less gain is needed. This technique will also bring up the bass a little due to the proximity effect of the mic, but the mixer EQ can be used to compensate for that. Next, you need PA and monitor speakers with a fairly flat frequency response and a nice even dispersion characteristic, as any significant spikes or humps in the frequency response will tend to set off feedback at those frequencies as you increase the gain. Similarly, any EQ boosts that might have been applied to the microphones increases the gain at specific frequencies, and more gain always equals a greater feedback risk. For the same reason it is always better to apply EQ cuts rather than boosts—turn down the bits you don't like, rather than boost the bits you do!

It is also pretty obvious that if your main PA speakers are pointing at the microphones, a lot more signal will find its way back into them than if the microphones are well behind the speakers. There are speakers, such as the columns made by Bose and other line-array products, that are designed to spread the sound over a very wide angle, and these can often be used behind the microphones, but the laws of physics still apply and you'll be able to get even more level if they're placed in front of the mics.

For the same reasons, monitor speakers pointing back at the performers can be particularly problematic with regard to feedback, but correct positioning can significantly improve the situation. Where cardioid mics are used, the insensitive rear of the microphone should be aimed directly into the monitor, whereas hyper-cardioid mics have their 'dead zone' somewhere between 30 and 45 degrees off the rear axis, as explained in Chapter 5. Check your mic handbook and look at the polar

Keeping your hand well clear of the basket prevents it from interfering with the directionality.

When the singer's hand cups the bottom of the basket, sound is prevented from reaching the rear of the capsule properly—an essential component of the way in which directionality is achieved in a cardioid mic.

diagram to see at what angle the best rejection occurs, and set the mic up accordingly. Where the singer intends to hand-hold the mic and walk about, they need to remember which part of the mic they're allowed to point at the monitors! They should also be warned not to hold the mic right up by the grille, as this will partially cover the ports at the rear of the capsule that control the mic's directivity, making the polar pattern more omnidirectional and feedback more likely.

Direct sound from the speakers isn't the only hazard—there's reflected sound to worry about as well—so if the stage has a reflective rear wall, as opposed to something absorbent like heavy curtains, and especially if it is close behind you, then your feedback worries have just got worse. A solid, low ceiling over the stage area also reflects sound, so making sure both your main-system and monitor speakers have a flat frequency response and well-controlled vertical dispersion will give you the best possible chance of using higher gains without provoking feedback.

As described in Chapter 1, column speakers and small line arrays differ from 'single woofer and horn speakers' in that the sound from the various drivers combines to reduce the vertical angle of dispersion while widening the horizontal coverage. There are other benefits too, such as better coverage towards the rear of the room, but in the fight against feedback, the fact that the sound is spread over a wider area means that reflections from surfaces that are directing sound back into the microphones are likely to be less intense, as their energy is spread more evenly through the room. Where the system has a separate sub-woofer, try to keep that as far away from the mics as possible to prevent rumbling, low-frequency feedback, and always engage the mixer's low-cut filter on all the vocal mics, especially if the sub is anywhere near the performers.

The inherent advantages of column speakers and small line arrays doesn't mean that you can't get good results from more conventional boxes, however, but you do have to position them carefully. A useful strategy is to place them on stands above the heads of audience members near the front (which both stops their bodies from soaking up all the sound and also avoids deafening them!), and then angle the speakers slightly inwards and downwards, so they aim at a point around two thirds of the way back in the audience. Some speakers, such as those made by HK Audio, have angled tilt-mounts built in, but if yours don't, it may be worth investing in stands that have tiltable heads.

'Ringing-out'

A very common setup procedure for PA systems is a process called 'ringing-out', which means turning up the gain on each mic until feedback just starts, then backing it off slightly. Once this has been done for all the mics individually you can check it again with them all turned up, and then

Speakers with angled tilt-mounts or stands with tiltable heads allow the sound to be more accurately directed at the audience.

TIP: Try to leave at least 5dB, and ideally 10dB, of fader headroom between your normal operating level and the point where feedback starts.

reduce the master level slightly if you hear the beginnings of ringing or feedback. Hopefully the acoustics will get better when a few bodies turn up to soak up some of the sound that might otherwise be reflected back, but you should nevertheless always try to leave at least 5 to 10dB of fader headroom between your operating level and the point where feedback starts.

Where feedback is particularly troublesome at just one or two particular frequencies, a third-octave graphic equaliser connected between the desk's outputs and the PA speakers may be used to reduce the gain slightly at those frequencies. You need to be aware, however, that the overall PA tonality will be slightly altered, as the graphic-EQ bands are far wider than the feedback frequencies they're trying to notch out. Graphic EQs are often used in the monitor feeds too, with the monitors being 'rung-out' in a similar way to the main speakers, first in isolation, to check everything is coming through OK, and then again with the front-of-house (main PA) turned up as well. The ringing-out procedure requires a degree of expertise to quickly match the feedback frequencies to the appropriate EQ bands, and to judge how the PA sound will be compromised by the process. I'd certainly advise against treating feedback with any graphic equaliser with bands wider than one third of an octave, as the tonal penalties are simply too great.

Automatic Feedback Suppression

Automatic anti-feedback devices are often used which are able to identify the precise feedback frequencies and then tune very narrow digital filters to exactly match those frequencies. These digital filters are usually just a fraction of a semitone wide, so their effect on the overall tonality is significantly reduced.

The usual way of using these devices is to turn up the system gain slowly until feedback occurs, wait for a filter to lock on to it and notch it out, then turn up the gain a little more until a new feedback frequency starts up, then wait for that to be notched out. If you do this for the first four or five feedback frequencies, you can leave any remaining filters to 'roam' so that they can pounce on any feedback that occurs during the performance, caused by, for example, a singer moving the mic around. Such systems can claw back useful headroom and also eliminate ringing when getting close to feedback—but they're not a 'cure' for bad practice in terms of speaker placement or inappropriate equipment.

Having tried a number of anti-feedback devices, I've found the majority of them work very well, although exactly how much extra gain you can get before feedback still depends on the room you are in and on the quality of the other components in your system. It's not unreasonable, however, to look for an additional 5 or 6dB of feedback-free gain, with an absence of 'near feedback' ringing.

INDIVIDUAL FEEDBACK PROCESSORS

System feedback suppressors are usually connected between the desk's main outputs and the PA amplifiers, and/or the monitor speakers in complex setups. Some personal-vocal-processing pedals also include automatic feedback suppression, and although these only provide a benefit to those individual performers using them, they will be effective on both the main and monitor speakers, as the filtering occurs at source.

Chapter Nine
Managing The Backline

In most small venues, a very significant part of the sound heard by the audience comes from the back-line—the amps and the drum kit on stage. The PA may only be needed to amplify the vocals and perhaps give the occasional leg-up for a guitar solo to achieve a well-balanced sound in the room, so keeping the back-line level under control and ensuring it produces the right sound is hugely important.

As a guitar player, I appreciate the sense of power that comes from playing through a large amp and speaker stack, but the reality is that your front-of-house mix engineer (or mixing musician, if controlling the mix from on-stage) will have a much better chance of producing a decent balance for the audience if the back-line levels are kept under control. No matter what the guitar player thinks (especially if he or she is wearing one of those Gibson T Shirts that says 'Vocals—wasted time between guitar solos'!), the band has to balance to the available vocal level, and not the other way around! I've seen and heard too many live-sound disasters where the guitar player has decided how loud they want to be without giving any consideration to anyone else. The overall out-front level of the PA and back-line combined also needs to set with a view to the level of sound that the audience will feel comfortable with (or that the venue is licensed for!).

Of course, if the venue is fitted with a volume limiter, as too many now are, you simply have no choice but to stay beneath its threshold level, or you'll constantly be having your power

Automatic volume limiters temporarily switch off the power when the volume in the venue exceeds a given threshold for a period of time.

turned off. If you are forced to play at a venue plugged into one of those awful things, try to ensure that any piece of gear that includes a computer or DSP section is plugged into a different circuit that isn't affected by the trip. When discussing this with the venue manager, make it very clear that this point isn't open to discussion, as, even where data corruption isn't a risk, there is still the problem that some digital devices (mixers, mains-powered computers, electronic drum kits, some modelling amplifiers and active speakers with internal DSP) can take half a minute or more to reboot, and if the whole show is going through a digital mixer, that could be a disaster. Try talking up the risk of both equipment damage and data corruption when pressing your argument.

For most line-ups, the audience will hear a better-balanced sound with at least some proportion of the on-stage instruments sound also coming from the main PA rather than just from the stage amps, so the practical question is about how to keep stage levels reasonable without compromising the performance. This is an especially difficult issue for those electric guitar players who rely on driving valve (tube) amps at a certain level to get their sound. And, of course, drummers!

Electric Guitars

From my own experience of playing guitar in bands, there are two main issues here, aside from that of being able to crank your amp hard enough to get the right sound: one is being able to hear the amplifier tone as the audience hears it, so you know what you sound like; and the other is being able to hear your instrument over the sound of the drum kit and other musicians. Stage monitors can help, but in small venues you may often only have vocal monitoring, so they certainly can't be relied upon.

Guitarists who rely on valve (tube) amplifiers tend to need to drive their amplifiers fairly hard to get classic rock tones, but a valve head and a 4 × 12 cab can be ridiculously loud, and even a cranked 20-watt valve combo can still be pretty antisocial! One option, if you really want to stay with a high-powered amp is to use a power-soak to reduce the level reaching the speakers. A properly designed power-soak poses no danger to the amplifier, and if you attenuate a 100-watt amp down to the equivalent volume of around a 15-watt amp you'll keep your essential tone, but the sound level will be a lot more manageable out front. Although, believe me, even that level can still be very loud in a small venue! This also brings with it, of course, a greatly reduced risk of hearing damage—a benefit that certainly shouldn't be underestimated.

An alternative, if you're still wedded to the idea of that big 4 × 12 cab, is to invest in a significantly lower-powered head, or perhaps one with a switchable-power output stage. The reality is that 15- to 20-watts of valve power (or 30- to 40-watts of solid-state power) will always be more than loud enough for a typical small venue, even without feeding it through the PA.

Small valve amps can sound wonderful when cranked up, as many classic recordings can confirm, and they will always produce a more convincing classic rock tone than using a 100-watt head with the master volume turned down. That's because the 'real' rock guitar sounds all rely on a degree of power-stage distortion, as well as preamp distortion, and master-volume controls reduce the level hitting the power stage, thereby preventing it from compressing and overdriving.

Speaker attenuators, like this THD Hotplate, reduce the level going to the loudspeaker, thus allowing valve guitar amps to be driven hard to achieve distortion and compression without producing excessive on-stage noise.

Of course, if you are happy with the sound of preamp or pedal distortion, you can ignore everything I've said so far and just turn your master volume or pedal output down!

So, a compact, low-powered combo is undoubtedly the most practical solution for gigging in small venues. It won't take up too much room on stage, it will fit into the back of your car, you won't feel it is trying to dislocate your shoulder when you carry it into the venue, and your PA engineer will have a fighting chance of balancing the band's complete sound. A good quality 1 × 12 or 2 × 10 combo will usually do the trick, and many pro players still use these even for festival-sized gigs, as they can get all the level they need via the PA and stage monitors. What matters is that you can get the sound that your want out of it and that you can hear yourself.

A small valve combo allows most guitar players to achieve a comfortable sound on stage without taking all control away from the front-of-house mixer.

Valves Are Not The Only Option

Even if you're a committed valve-amp user, it is worth trying out some of the current crop of modelling or hybrid amplifiers, as you might be surprised at how good they actually sound. Their big advantage is that they are able to sound consistent over a wide range of volume levels i.e. they don't have to be loud before they give of their best. They also tend to include useful effects such as reverb, delay and often modulation too, which not only saves on the cost of pedals but also allows setups for specific songs to be saved and recalled as presets. Hybrid amps combine old-school, analogue, valve circuitry with DSP wizardry, while modelling amps are all DSP, but most have in common that the final sound is amplified via a clean amplification stage, giving you full control over volume without the tone changing.

Other plus points include lower initial cost and reduced weight, compared with their all-valve equivalents. There's also the issue that valve amps change character as the valves wear out, and replacing a set of valves every year is a fairly expensive business. Modelling amps may never satisfy those with years of experience playing through valve amps, but the better ones can get very close to the sound and, just as importantly, the 'feel' of playing the real thing.

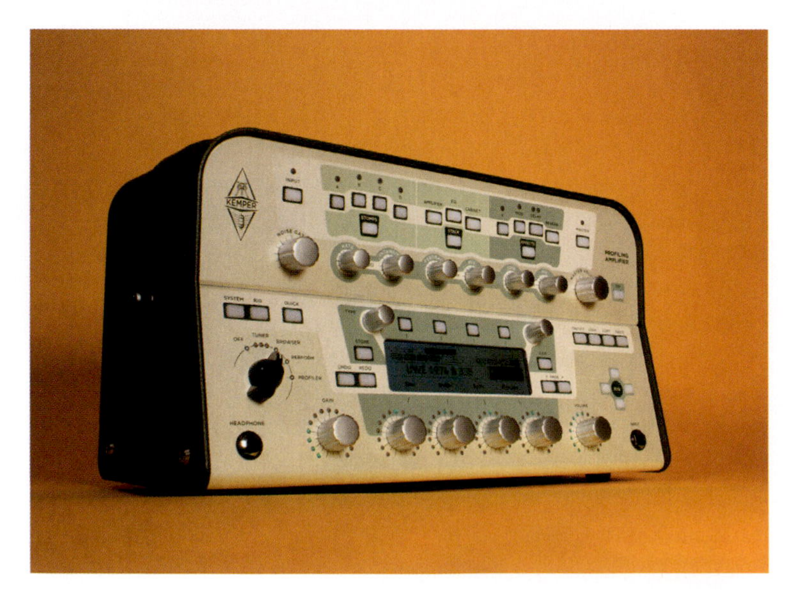

Digital modelling of guitar amps has reached the point where many players are now happy to use it in place of tube amps. This Blackstar amp comes in traditional combo form, whilst the innovative Kemper Profiling Amp can be plugged straight into the PA.

Some players choose to dispense entirely with traditional guitar amplifiers and instead use a digital-modelling preamp plugged directly into the PA, relying on stage monitors to hear what they are doing. If the guitar player is happy with the sound, this approach makes the mix engineer's job much easier, but all the control is in the hands of the guy out front, so the player certainly has to have faith in whoever is mixing the gig! With this kind of setup, it is crucial that the monitor level is just right for the performer, so a personal monitor that allows independent adjustment of volume is a good idea.

The Axis Of Tone

If you prefer to be able to hear your own amp and speaker on stage, rather than your miked-up sound coming back over monitors, you need to be very aware that the tonality changes dramatically as you move away from the frontal axis of the speaker. It is all too common to see a guitar player with an amp on the floor pointing at their knees. The practical outcome is that the player doesn't hear the full volume of the amp, or most of the high frequencies. The inevitable reaction is to wind up the volume and add more treble, until the tone and the balance seems right from their playing position. Meanwhile, the front rows of the audience get their ears sliced off by the volume and extra top end.

A simple way to avoid this scenario is to buy a low stand that allows the amp to be angled steeply so that the speaker is pointing towards the player's head—or, alternatively, to use a taller amp stand that achieves the same result.

Some players even choose to place their combo amp out in front, facing back towards them, and angled up in the style of a floor monitor. This has the dual advantages that the player gets plenty of level and hears their tonality accurately, whilst little of the sound makes its way directly to the audience, thus giving the mix engineer much more control over the balance of the band. It also ensures that everyone else hears some guitar without needing to put as much in the monitors. Furthermore, if the player also sings, this arrangement will reduce the amp spill being picked up in their vocal mic.

A low tilt stand on the floor helps to get your sound up around your ears, rather than firing it into your back!

Overdrive pedals can be used to preset an appropriate level boost for guitar solos, keeping the performers more in control of the musical aspects of the balance.

Guitar players often need some means of boosting their level for solos if they don't want to rely on the PA engineer doing it for them. Many old-school players simply use the volume control on their guitars—with the right amp or FX pedal, pushing up the volume increases both the level and amount of overdrive—and this can be an extremely intuitive way to work. Alternatively, an overdrive pedal that adds volume, with only a little extra dirt, such as a carefully set Tube Screamer type, or a Fulltone OCD, is a good-sounding and practical option. With modelling amps, the user can switch presets, using a footswitch, to an appropriate solo setting.

Bass Instruments

Electric-bass guitar players don't rely on distortion to achieve their sound, so with no amp-volume 'sweet spot' there is usually more flexibility in setting the on-stage level. It is also fairly common now for bass guitar amps to include a balanced XLR DI output, so getting signal into the PA is straightforward. Today's sophisticated compact bass combos can be very loud, with well-designed cabinets offering plenty of tonal depth, so even a single 12-inch speaker may often be able to do the job that used to require a 4 × 12 cabinet or a 2 × 15-inch speaker system. Low bass notes invariably cause the snare drum to rattle, and while increasing the distance between the bass amp and drum kit may sometimes help a little, it's a problem we all simply have to live with on small stages.

Acoustic Instruments In An Electric World

It is very difficult to amplify acoustic instruments in the context of an electric band using just microphones. You simply cannot get enough level before feedback. DI'ing the instrument via a pickup solves that problem, but usually at the expense of achieving a 'natural' sound. In the context of a loud band, some form of monitoring will also always be needed if the performer is to hear what they're playing. At larger venues, conventional wedge or side-fill monitoring (or an in-ear system) will work fine, but at smaller gigs, where there isn't even space for an acoustic combo, you could utilise a compact, active personal monitor.

Amplifying acoustic instruments alongside loud electric instruments is always a serious challenge.

These can be mounted on a mic stand—sometimes you can just split your usual boom stand at the pivot point and mount the monitor in the middle, saving even more space—and you simply plug the instrument directly into the monitor, with a thru feed to the PA. Heard in isolation, personal monitors project very little low end, but you'll always be able to hear plenty of bass from the main PA, as low frequencies radiate almost omnidirectionally.

Where there is space on stage for a dedicated acoustic-instrument combo, the same considerations apply as with the electric guitar: the performer needs to be able to hear what they are playing with an accurate representation of the tonality, and the level on stage must be low enough to give the mix engineer enough scope to achieve the desired overall balance for the

Personal vocal monitors intended for mounting on a mic stand can work well for acoustic instruments too.

audience. Unlike most electric guitar amplifiers, acoustic instrument amplifiers are designed to produce as clean a sound as possible across their entire volume range. Many will also include an adjustable notch filter that can help tackle the main feedback frequency: usually the primary resonant frequency of the instrument's body.

All-acoustic Acts

All-acoustic acts inherently tend to have a more natural internal balance than electric bands, and even where acoustic combos and bass amps are used, the on-stage levels usually remain quite modest. Furthermore, all-acoustic acts don't generally require a lot of out-front volume, unless they are playing in a very large venue, so the PA just needs to provide a comfortable listening level with good vocal clarity.

Without loud drums or stage amps to compete with, you may be able to mic the instruments and benefit from the natural sound, but where the instruments have pickups, it is always worth bringing those into the mixer, so you can use them to rectify any particular mic-capture issues that might arise during the performance.

Mainly acoustic bands are often best tackled with a combination of solutions: instrument-mounted mics, pickups and conventional miking.

Violins are much easier to handle if they are fitted with a pickup, as the performers invariably move quite a lot, and the best-sounding position from which to mic them is a couple of feet or so above the instrument. In a live situation this just invites huge spill problems, even if you can manage to get enough gain before feedback. In general, though, normal miking practice can be followed, with a mic position and distance appropriate for the instrument, avoiding as much as possible any nearby sources of spill or reflective surfaces—see Chapter 11 for more details on this topic.

Accordions and melodions are sometimes fitted with internal mics, but if not, they can be picked up using one mic on each side (one for the bass notes, the other for chords and melody), but these do have to be kept at a reasonable distance because of the movement of the bellows. I find that a spacing of one to two feet either side is usually OK, and I have successfully used omnis rather than cardioids in this application. Cardioids may produce less spill, but what spill there is can sound quite boxy because of their uneven off-axis response, so a greater amount of cleaner-sounding spill will often compromise the overall sound quality less.

When miking an all-acoustic act, I tend to place the monitors further from the performers than I would for electric acts (especially if any omni mics are being used), so that the coverage is more even for the individual players. In some cases cross-stage side-fills are enough, although I'll often add a centre monitor too on wide stages or if the main performer needs a different monitor balance. I'll also try to keep the main PA speakers as far from the musicians as possible, to minimise feedback, and I'm also very careful, when ringing-out the system, to ensure there's plenty of headroom before feedback. In a typical 200-seat venue, I normally use a 1.5 kilowatt compact line-array system, augmented by a single 12- or 15-inch sub, and running nowhere near full power to give plenty of level as well as excellent clarity.

Drums

While amplifiers can be turned up or down, acoustic drum kits have no such luxury. In smaller venues, the main culprit, when it comes to excessive volume, is always the snare drum. This is often the source that determines the level everyone else feels they have to play at. Using damping pads or gel dampers will change the decay time and brightness of the drum slightly, but tends to do little to reduce the sound level, and I haven't yet come across a snare attenuation system that doesn't also significantly change the sound (Ringo's tea towel, anyone?). It's also true that some snare drums are significantly louder than others, so it's a good idea for drummers to have a quieter model for use at smaller gigs. Hiding the drummer's ear plugs is another way to ensure a more restrained performance!

For large-scale concerts where sound quality is at a premium, it is becoming increasingly common to enclose the drum kit within heavy Perspex screens, to reduce spill, from the kit into the other mics and vice versa. Of course, this isn't a practical option on the small-venue circuit, but if you ever get the chance to use them, they are really effective at cleaning up the on-stage sound.

While it may not seem very macho to use 'quiet' drum sticks, these often do a good job without seriously compromising the

The snare drum is invariably the loudest source on stage and one of the most difficult to control without compromising its sound and playing feel.

sound. You'll find several styles available, although the ones made from a bunch of smaller diameter rods seem to offer the best balance of volume, sound and playing feel.

The most radical solution is to use an electronic kit, but the drummer needs to be comfortable playing on one. While electronic kits fitted with the latest mesh heads have a good playing feel, they can also be expensive compared to a typical acoustic kit, especially when you take into account the cost of suitable amplification as well. They can certainly sound very authentic though, and can be mixed at any level the venue requires.

If you need *really* quiet drums, 'quiet' drum sticks like these Hot Rods may be the only answer.

SELF-BALANCING FROM THE STAGE

Where you have to balance the band without the benefit of a front-of-house engineer, and you need to hear how your guitar sounds, a long instrument cable or radio system will enable you to go out-front during your sound check. In practice, you won't be able to go too far whilst still playing before the delay due to the speed of sound starts to make it difficult to stay in time with the rest of the band, but in most venues, the out-front sound is significantly different from the sound you hear from the stage, so it is essential that at least one of the band members is able to evaluate it.

Backline Spill

A lot of back-line sound usually leaks into the mix via the vocal mics, so it is always worth trying to move the mics and amps around to minimise spill. In very cramped venues, singers will often stand directly in front of the drum kit, which can result in almost as much drums as vocals coming through the vocal mic. Moving the vocal position a little to the side of the kit can make a useful difference, and the same is true of guitar amps that are aimed directly at vocal mics. This becomes even more of an issue when the act includes miked acoustic instruments that aren't going to be playing all the time. A mix engineer would fade unused mics up and down, as needed, but when you are mixing from the stage, the chances are that the mic will be on all the time, so it's worth taking special care with the positioning to minimise spill.

There is always a temptation for performers to turn up their instrument volume as the gig progresses as the human hearing system automatically adjusts in response to continual loud sound, so that it no longer seems loud to them. However, once a balance has been achieved during the soundcheck it is very important that individual band members don't change anything as there's no way of verifying the balance is still OK without someone going out front again.

Once a good balance has been achieved, it helps if the performers don't alter their on-stage volume too much.

Hearing Damage

Which leads us neatly on to our next topic—it would be irresponsible of me to ignore the subject of potential hearing damage caused by exposure to loud sounds over an extended period. Individual tolerance varies, but there are official 'noise dosage' guidelines that make sense, and some smartphone apps are starting to appear that measure noise exposure during the soundcheck. The numbers you're looking for are calibrated in LAeq, not SPL!

It is generally stipulated that an individual's maximum noise 'dosage' should be restricted to an average sound level of 85dB LAeq over an eight-hour period, with the 'safe' period halving for every additional 3dB of level. In practice, this means that, for a typical two-hour gig, the noise exposure should be no more than 91dB LAeq. Don't confuse these LAeq figures with peak

Making custom moulds for in-ear attenuators. Off-the-shelf products are better than nothing, but if you are serious about looking after your hearing, custom moulds will do the best job.

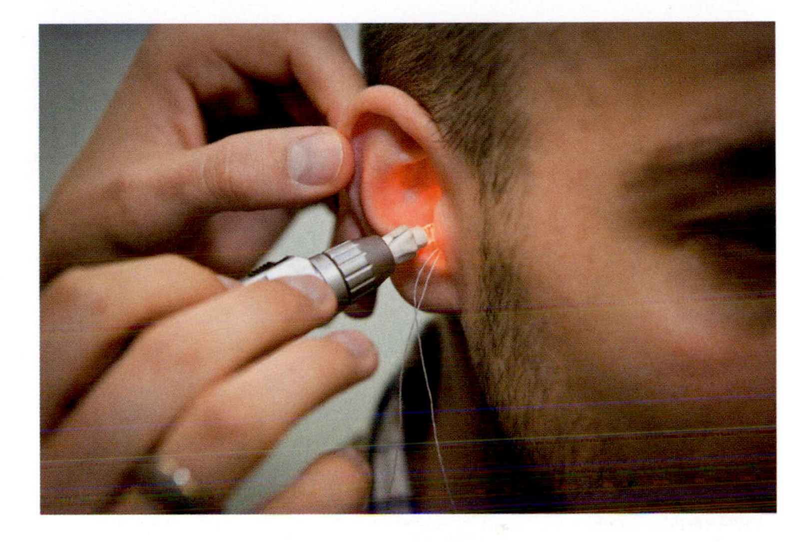

SPLs, though—music naturally has repetitive high SPL peaks set within a much lower average sound level, and noise exposure is all about the average level over long time periods. A 91dB LAeq exposure is actually pretty loud! Nevertheless, given that the on-stage volume, close to the amplifiers and drums, will usually be significantly higher than that which the audience is exposed to, you can see that this is a serious issue for musicians. Any form of ear protection will help avoid hearing damage to some extent, but it can also interfere with your ability to play an instrument or pitch vocals properly, and can actually encourage some musicians to play even louder, thereby worsening the situation for everyone else! Custom-moulded earpieces, which can be fitted with 10 or 20dB 'flat-response' attenuators, are an excellent, though pricey, option, as they retain something close to a normal tonal balance while also providing very good protection.

Chapter Ten
Radio Microphones And Guitar Systems

Radio mics and guitar transmitter packs have been used by gigging musicians for many years now, although the quality can vary substantially depending on the manufacturer, the technology, budget, and the operating frequency. The last point is very important, and a major source of confusion since some radio channels are only available to licensed users, and some are free for everyone to fight over! Frustratingly, many of the licensed and free radio frequencies changed in the UK in 2012 as part of the 'digital TV switchover' process, making a lot of older radio mics and guitar systems, which can't be retuned, obsolete and illegal.

Spectrum allocation and licensing requirements differ all around the world, so I suggest you consult with your dealer or the relevant licensing authority in your territory before investing in wireless technology.

Modern wireless systems offer audio performance comparable to a conventional wired signal, but with the added advantage of greater mobility.

WIRELESS SPECTRUM CHANGES IN THE UK

The analogue to digital TV switchover in the UK has freed up large parts of the broadcast frequency spectrum, which the government has sold to providers of services such as 4G wireless broadband and mobile phones. Parts of the sold spectrum included the airspace previously used for licence-free radio mics, and hence the need either to re-tune existing equipment to the new channels, or purchase new equipment designed for the new channel allocations. Currently it is a legal requirement in the UK and Europe to have a licence to operate radio microphones on channels other than those in the de-regulated part of the spectrum in CH 70 (863–865MHz). If you are in the UK and have any concerns over your existing equipment or channel licensing, I recommend contacting the Joint Frequency Management Group (JFMG) on www.jfmg.co.uk.

In the UHF part of the spectrum we have free access only to the 863.100 to 864.900 MHz region (CH 70), where the available transmitter frequencies are: 863.1, 863.3, 863.5, 863.7, 863.9, 864.1, 864.3, 864.5, 864.7 and 864.9 MHz. Adjacent bands can interfere with each other so it is safer to set them at least one frequency-band apart, or even two if possible, which may mean you can only use three devices at the same time. The interference is due to intermodulation, where two or more frequencies combine to produce additional sum and difference frequencies. In professional productions such as stage musicals, where many radio mics may be required, great care is taken to calculate compatible transmitter frequencies which are spread across many location-specific licensed channels.

Of course the '802.11' wi-fi band used by our internet routers is also licence free, with up to 70 usable frequencies at the same time within the 2.400 to 2.4835 GHz range. While this is rarely used for professional wireless microphones due to fears of interference from consumer wi-fi, Bluetooth, car keyfobs and even microwave ovens, a number of digital radio mic and guitar-pack systems that work on these channels are available for the gigging musician. The ones I've tested so far have worked very reliably with good sound quality, very low noise, low distortion and a wide dynamic range.

Digital Radio Mics

So-called 'hybrid' digital wireless systems use FM transmission in combination with DSP technology to reduce noise and to flatten the frequency response. In other words, these use what is essentially an analogue system with a little DSP-help to optimise its performance. By contrast, a purely digital system usually employs a form of data-reduction, similar in principle to that used in MP3 files, which reduces the radio-frequency bandwidth needed to carry the signal. An RF carrier is encoded with the digital audio stream, with various error correction strategies to compensate for small errors due to data loss or momentary interference. Purely digital systems have an inherently wider dynamic range than typical analogue systems, but do impose additional signal processing delays of a few milliseconds. Encryption can also be included to prevent 'electronic eavesdropping', which can be important in conferences and other non-music applications.

Although rarely used for professional audio applications, there are now a number of digital radio mic and guitar-pack systems that utilise the 2.400 to 2.4835 GHz wi-fi spectrum, offering good sound quality with very low noise, and distortion.

TIP: Radio systems work most reliably when the transmitters and receivers have a clear 'line of sight' between them—any significant physical obstructions have the potential to reduce both range and reliability.

Digital systems designed for use in the 2.4GHz wi-fi band automatically select the most reliable frequency, and as they can use metadata to prevent other sources being misinterpreted as valid signals, they can still work very reliably even when close to internet routers or audience members with wi-fi active on their mobile phones.

Effective Use

Radio systems always work most reliably when the transmitters and receivers are in 'line of sight' of each other—any obstructions have the potential to reduce both range and reliability. Where a large number of radio systems are to be used at the same time, as in the case of a large-scale musical theatre production, then the only solution is to license some of the additional Band IV (TV) spectrum that can be licensed for special events.

TIP: The fewer radio devices you are trying to use at the same time, the less likely you are to encounter intermodulation issues, but there is still a chance that you may experience problems resulting from other people using the same parts of the unlicensed spectrum near your location.

As a rule, though, the fewer radio devices you use at the same time, the less likely you are to experience intermodulation problems, although there is still the possibility of other people using the same parts of the unlicensed spectrum near your location. One big plus of using wireless systems in smaller venues rather than large arenas is that the transmitter and receiver will always be fairly close to each other, which helps significantly in avoiding interference from more distant sources.

The type of receiver also makes a difference to operational reliability, with the most effective type being known as 'true diversity'. As transmitters are moved around, radio-wave reflections from walls or metal frames cause phase cancellation that may result in signal drop-outs. Non-diversity receivers with only a single antenna are at risk of such dropouts, but a true diversity receiver has two separate receiver circuits fed from two sets of antennae. When one antenna is in a dead spot, the other will probably still be picking up a viable signal, and a comparator circuit inside seamlessly switches to whichever antenna/receiver has the strongest signal at any given moment. A less sophisticated type of diversity receiver uses twin antennae but a single

receiver circuit, but again a switching system takes the feed from whichever antenna is picking up the best signal.

A 'true diversity' receiver uses two antenae and two receiver circuits, switching seamlessly between them to maintain the strongest signal at all times.

Battery Power

Transmitter packs are powered by batteries, to allow them to be fully mobile, and whilst fresh batteries can give several hours of reliable operation, when the batteries near the end of their life, their failure can be sudden and dramatic (especially with

At any level of performance, it is worth fitting new batteries for each show. Partially used batteries decline unpredictably and failure can be sudden.

systems that use special circuitry to squeeze every last bit of life out of the battery). At any serious level of performance, it has to be worth routinely fitting fresh batteries for every gig, as the pros do. Some systems can work with rechargeable batteries, although the operating life may be shorter than for dry batteries, depending on the type of rechargeable cell.

Many radio transmitters include a battery-condition indicator, but as these may be impossible to see during performance the better systems transmit battery status to a display on the receiver. Remembering to turn off the transmitter pack during breaks and when you've finished using it, avoids battery wastage and embarrassing comments being broadcast without your knowledge! At least one system I've tested automatically powers down the transmitters when the receiver is switched off, which certainly seems like a good idea.

Transmitter And Receiver Placement

Receivers for the radio mic systems are usually placed on stage, as close as is reasonable to the performance area, but not so close that a transmitter is likely to get within a metre or so of it —some systems can become overloaded and misbehave if the transmitter and receiver get too close. It's a good idea to test your own system to see how close you can get before problems occur.

Whilst a line of sight should ideally always exist between the performer and receiver, where you can be certain that the performer will never stray too far, the occasional obstacle between the transmitter and receiver may be acceptable. It is good practice to establish the maximum reliable range prior to the sound check, and make sure that the performers are aware of how far they can safely move.

Old VHF radio systems lost efficiency if the transmitter's trailing antennae wire was too close to the performer's body, but while modern systems seem rather better in that respect, it's still best to place the transmitter pack over at least one layer of clothing rather than directly against the skin and to have any flexible antennae bending away from the body rather than towards it.

Whilst the occasional obstacle between the transmitter and receiver is usually acceptable, you should plan to maintain a line of sight between the performer and receiver at all times.

Guitar Systems

The first-generation of wireless systems for guitars were notorious for compromising the tone and feel, as well as being excessively noisy. In particular, they couldn't handle large changes in level, such as would be encountered with a guitar player 'riding' their volume control as a normal part of performance. Newer digital systems leave all those shortcomings behind, but one thing many designers still fail to

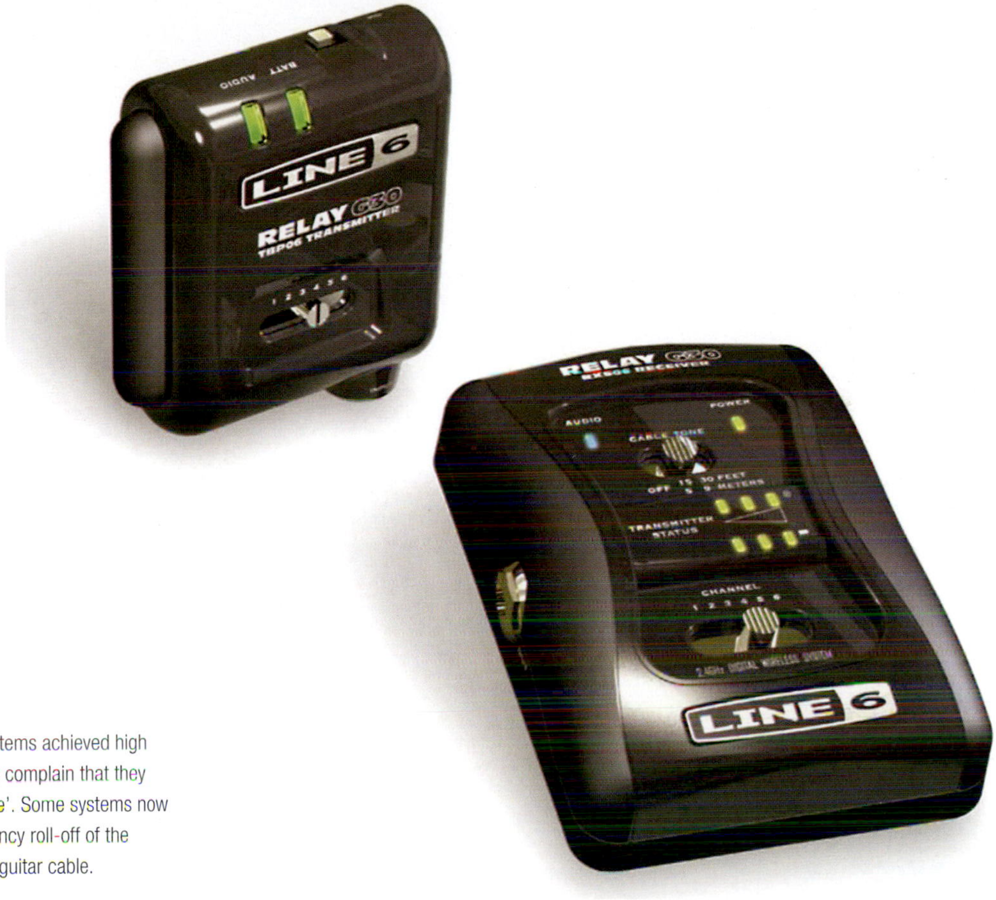

Once guitar wireless systems achieved high quality, players began to complain that they didn't 'sound like a cable'. Some systems now emulate the high-frequency roll-off of the capacitance of a typical guitar cable.

take into account is that the capacitance of a typical guitar cable is an integral part of the sound, and using a short cable to connect to the radio pack reduces the capacitive loading, thereby making the sound unnaturally bright.

Consequently, some guitar-link radio systems have switchable 'cable emulation', but to work correctly these must be on the input to the transmitter, not on the receiver, as the cable capacitance interacts directly with the inductance of passive guitar pickups to subtly modify their tone. In an ideal world the transmitter input would have an impedance comparable with that of a typical guitar amplifier as seen through a typical guitar cable—most have a suitably high input impedance, but the ideal cable capacitance is a subjective thing, as different brands and lengths of cable exhibit different values of capacitance, so a simple means of fine-tuning the capacitance would be the ideal option.

To meet precisely this need, connector-manufacturer Neutrik have come up with a jack plug with switchable capacitance to allow a short guitar cable to emulate the capacitance of longer ones. You can switch between the different capacitor values while playing and then leave it fixed on whichever one sounds the best.

You can, of course, make up your own cable by soldering the appropriate value of capacitor across the jack terminals inside the amp-end of the cable. A typical guitar cable has a capacitance of around 100 picofarads per metre, so simply multiply that value by the length of cable you normally use to arrive at an equivalent capacitance. You may need to experiment with the final value to find the one that sounds best.

An alternative approach to tuning a wireless guitar system to sound like a cable involves adding capacitance to the short length of cable between the guitar and the transmitter. The Timbre plug from Neutrik is a particularly elegant solution.

A guitar system receiver would normally sit on top of the amp, ideally with a Velcro strap or something similar to prevent it getting accidentally knocked off. If you use a pedalboard, however, the receiver will need to go before the first pedal in the chain, so it will have to be fixed to your pedalboard. Many receivers will run from a Boss-compatible nine-volt power supply, so you may be able to run it from your usual pedalboard PSU—just make sure it can supply enough current. Obviously you need to place the receiver where you won't tread on it, which probably means close to the back of the pedalboard, and bear in mind, too, that the space around the antennae needs to be kept free of significant obstructions and cables.

The final critical aspect of setting up a guitar or voice wireless system is the adjustment of the transmitter's gain control to avoid overloading the transmitter—and it's better to err on the low side, if in doubt. For guitar systems, the receiver's output level may also be adjustable to match the amplifier's or pedal's input sensitivity.

Chapter Eleven Microphone Choice And Placement

At smaller gigs, it is often not necessary to mic-up the back-line at all, as it may well be quite loud enough already, but for larger venues, where the audience will predominantly hear the sound from the PA, it is important to know which mics to use and where to position them. This chapter discusses which microphone types or DIs are best for the most common on-stage tasks, as well as microphone placement and initial EQ settings.

The principles of microphone operation were covered in Chapter 5, but when it comes to selecting specific models of microphone for live use there are several things to take into consideration, not least being budget. While very cheap microphones are invariably a false economy, there are many highly affordable models on the market now that turn in a good performance. Most mics designed for stage use will be reasonably robust and, if properly looked after, should last a lifetime, so it makes sense to buy the best you can afford.

One way to make sure your mics continue to give of their best year after year is to make sure they are securely mounted on good-quality mic stands. The brittle plastic type of stand clip that seems to come with many mics tends to develop fractures, so always inspect them before entrusting your mics to them—the slightly rubbery, more flexible ones are tougher and less prone to sudden failure. If stand-borne noise is a problem, take a look at Rycote's InVision shock-mounts, as they are virtually unbreakable and visually unobtrusive. Unless

➤

Mics designed for stage use are robust and should last a lifetime if they are properly looked after.

➤

Always tape or clip the mic cable to the stand. Sometimes it helps to leave a loop at the mic end to kill any vibrations travelling along the cable itself.

the mic will have to be removed during the performance, then always tape or clip the mic cable to the stand to help kill any vibrations travelling along the cable, but leave a loop at the mic end so that you can angle the mic easily. Always position boom stands with the boom arm in line with one of the three legs for maximum stability, especially when using heavier microphones. Tall stands with long boom arms, such as those used for drum overheads, may require a sandbag or stage-weight across the base for additional safety.

PHANTOM-TOLERANT . . . OR NOT?

Many smaller mixers have their phantom power switched globally, so it is very important that you check any amplifiers and instrument preamps fitted with DI outputs to confirm that they are able to tolerate phantom power being applied to their outputs, even though they may not need it to operate. Most properly designed equipment is happy to tolerate phantom power, though I've come across a few acoustic instrument preamps systems that won't work correctly, and some that could be damaged by phantom power. If you come across any of these, the only practical option is to route them via a conventional DI box to isolate them from the phantom power source.

Vocals

As discussed in Chapter 5, the usual approach, certainly in the pop/rock music genre, is to use a dynamic, cardioid microphone as close to the singer as possible, so as to both maximise level and reduce spill. A model designed for vocals will usually have a presence peak, to help the singer cut through, and a rolled-off low end to reduce the severity of the proximity effect when used up-close.

Although standalone pop shields are almost always used in the studio, they are not practical in a live situation, and the foam windshield often supplied with mics do little to combat vocal 'plosives'. Popping can, to some extent, be controlled by the singer through backing off or singing slightly to one side of the mic when pronouncing Bs and Ps, but using a mic with an integral low-frequency roll-off and/or engaging a low-cut switch on the mixer will help considerably.

TIP: It is always worth trying a few different mics to see which one suits the character of a singer's voice best—this seems to be especially important with female vocalists as their generally higher register can show up undesirable mic characteristics that may not be significant to a male singer.

The shape and location of the presence peaks in a mic's frequency response make quite a difference to how the high-end sounds, and my aim is always to try to find something that has good intelligibility and a nice airy quality, without sounding strident. You really need to test vocal mics with your own PA system, as the final sound reaching the audience is a combination of the tonal character of the speakers as well as

All mics intended for on-stage vocals will incorporate an integral pop shield, but a vocalist with a good mic technique can still make an improvement by laying back on 'plosives'.

the microphones. Capacitor mics suitable for hand-held vocal use are becoming increasingly prevalent, and offer a more open, natural-sounding high-end than dynamic models, but you need to have good PA speakers, operating in a good acoustic environment, before most people will appreciate the difference.

Wind Noise

While the foam windshield that comes with most vocal mics is of very limited use against vocal popping, you shouldn't lose them or throw them away. Wind noise can cause real problems when doing outside gigs, and foam windshields on the mics may make the difference between being able to work and having the wind howling louder than the singer! Even relatively small gusts of wind can sound very loud when picked up by an unprotected microphone. I keep a pack of coloured windshields in my kit bag for small outdoor events—having a different coloured 'sock' on each mic helps you keep track of them if they get moved around between bands. Where possible I'll also fit a matching-colour mic cable, again to make individual mics easier to identify.

Electric Guitars

The sound from a guitar amp changes significantly depending on the position of the mic relative to the speaker. For live work the mic always needs to be fairly near to the speaker grille to minimise unwanted spill from other loud sources on stage. The brightest sound is to be found with the mic pointing directly towards the centre of the speaker cone, and as you move the mic towards the edge, or if you angle the microphone, the sound tends to become more mellow. Thus, you can use the mic positioning to tune the sound to a large extent before even thinking of reaching for the desk's EQ.

The punchy midrange of dynamic mics makes them a favourite for guitar-amp work, although some players prefer the greater transparency and cleaner top end you get from a capacitor microphone. Some brave souls even use ribbon mics on stage—these can produce great guitar sounds, with throaty mids and highs that are smooth without being dull. However, you must remember to treat ribbon mics with great care. Some of the modern ones are quite robust, but they still need to be treated with a certain amount of respect, as dropping them or allowing a mic stand to topple over could break the ribbon.

TIP: Try not to 'hot-plug' ribbon mics into a phantom-powered cable in case the hot and cold pins don't mate simultaneously. Most ribbons will tolerate phantom power but you should always plug the mic in first and then switch on the phantom power.

For all but the smallest of guitar amplifiers, a mic that can handle at least 125 to 130dB SPL is essential, as the volume close to a guitar-amp speaker grille can get very loud indeed. You'd also be surprised how much the choice of microphone affects the final sound and, as a guitar sound is much more about art than science, the mic with the best technical spec won't necessarily give you the best sound. It really is worth trying out all the mics you have (or can borrow) to see what works best on any particular guitar amp. That said, you can pretty much rely on a tried and trusted choice such as a Shure SM57 to give usable results in most situations. Commercial sound companies often use a lot of SM57s as they are tough, not too costly, and give

➤
You can use mic positioning to tune the sound of a guitar speaker to a large extent before even thinking of reaching for the EQ.

good results on guitar amps, close-miked drums, and can also be pressed into service for miking brass and percussion, or even vocals if necessary. They may not be the 'state-of-the-art' in any single application, but you can always count on them to give you something you can work with. Another very popular all-rounder, costing a little more, is Sennheiser's MD421—a hugely versatile dynamic mic that works on just about everything, from vocals to kick drums. If you find yourself on a desert island that happens to be equipped with an XLR input, the MD421 would be a good mic to have on your side!

You may sometimes see a mic hanging down by its lead in front of the guitar amp. This may mean that the band or PA provider has run out of mic stands, but it just might also be for valid reasons. It doesn't matter that the mic's most sensitive axis will be pointing directly at the floor, rather than at the speaker. A cardioid mic is only 3 to 6dB less sensitive to sound from

90 degrees so there will still be enough level. However, this approach employs the full extent of the high-frequency roll-off exhibited by the mic's off-axis response, which can sometimes be the perfect 'pre-EQ' for taking the harsh edge off an overly-aggressive rock guitar sound, especially the sort you often get when the player stands off-axis from their amp and then turns up the treble to compensate.

If you are using the 'hanging method' to avoid using a mic stand, rather than as a tonal choice, then consider using an omnidirectional mic instead because its off-axis performance will be almost identical to its on-axis response, and spill will be virtually identical to a cardioid in the same position.

Utilising the high-frequency roll-off exhibited by the mic's off-axis response can sometimes be the perfect 'pre-EQ' for taking the harsh edge off an overly-aggressive rock guitar sound.

Do, however, take care to prevent the mic rattling against the grille of the amplifier, especially if the grille is made from perforated metal. This is another situation where a foam windshield can come in handy, as the foam provides a soft cushion between the end of the mic and the speaker grille.

More conventional miking, whilst still avoiding the clutter of additional mic stands on tight stages, can be achieved using mic-mounts that clip onto the guitar cabinet itself. If you are using a conventional mic stand, stage rumble can enter the mic via the stand, just as it can with vocal mics, so it may still be prudent to switch in the low-cut filter on the mixing desk. Although this will thin out the low end slightly, so long as it is a cardioid model, the proximity effect from using the mic very close to the speaker will more than make up for it. Where there's more than one speaker in a cabinet, it is always worthwhile taking a moment to establish which one sounds best—they will often be slightly different, and that can give you further miking options.

Speaker Simulators

In Chapter 6 I explained that a speaker emulator is a filter circuit designed to replicate the characteristic frequency response and distinctive steep high-frequency roll-off of a typical guitar speaker. There are two distinctly different types: those designed to be connected in line with a loudspeaker, and those intended to to be used with a preamp output. Speaker-level signals are of a much higher voltage than line-level signals and so should only be fed into DI boxes, preamps or speaker simulators specifically designed to accept speaker level signals. Some speaker simulators include a dummy load and can safely take the output of an amplifier without a speaker being connected. A dummy load will be clearly identified as such and, just like a real speaker, its impedance and power rating needs to be properly observed. Most players still want to hear their amp on stage so a DI/Simulator device with a thru connection is the most common option, but one situation where you may choose to use a combined speaker emulator and dummy load is where the amplifier is simply too loud to use conventionally.

Simulators designed for speaker-level signals will also feature a 'speaker thru' jack, to allow a speaker to be run at the same time. Where the simulator also incorporates a dummy load and attenuator, the combined impedance must be taken into consideration.

A line-level-input speaker simulator is easier to rig, as it can connect to any preamp or slave output on the amp, but may lack the sonic contribution of the amp's output stage, depending on where in the amp's signal path the output is actually taken from. The output from a speaker DI/Simulator box may be at mic or line level, depending on the model, but should be balanced. The most convenient type for live use will have a balanced mic-level output on an XLR connector, which can be connected directly to the stage box and treated just like any other mic source.

In most instances, there's actually little to be gained from DI'ing a guitar amp rather than miking it, as it is not difficult to get a usable miked sound, and the lack of a direct electrical connection between the mic and amp means that ground-loop hum won't be an issue. One exception is guitarists using preamps, such as a Line 6 POD or Fractal Audio AxeFX, which are designed to be fed directly into a full-range speaker system. These usually have stereo outputs, often on line-level XLRs, so watch out for clipping your mic inputs if you connect them directly to your stage box. Better to use a DI box, which may give you ground-lift options too.

Bass Guitar

Bass guitars are frequently DI'd in preference to miking, and there are some excellent dedicated bass preamps available for those who want the convenience of a DI but the sound of a real amp. Most high-quality bass combos and heads now come with a balanced, post-preamp DI output. Sometimes, a direct feed from the instrument is still useful, though, to avoid having the PA-feed compromised by the player's preferred on-stage EQ setting. Bass guitars fitted with standard passive pickups should be mated with a high-impedance DI box (usually an active model) to avoid loss of both high-end and level.

Bass speaker cabs seem to benefit from being miked a few inches further back than you might choose for a guitar speaker, but make sure you know how much of the bottom end is being rolled off by the absence of proximity effect—some models exhibit dramatically more than others.

Active basses (and guitars) have buffered outputs that can be plugged into passive or active DI boxes.

Miking a bass amplification system is similar to miking guitar speakers, the main difference being that the sound often benefits from the mic being set up a few inches back from the speaker grill, rather than being pressed right up against it. It is also self-evidently important to choose a mic that has a good bass response, rather than a vocal mic, where the bottom-end will be rolled off to counteract proximity effect. In most cases low-cut filtering will not be required, but let your ears decide—if you are working with a PA system that doesn't have a sub, you may need to engage a bit of low-cut to moderate the level of deep bass fed through the full-range speakers to avoid overtaxing them.

A little broad-band EQ may be needed to achieve a more even response across the range of the instrument, and to deal with room resonances. 'Modern' bass-playing styles will also often benefit from compression, particularly because the different techniques—plucking, slapping and pulling—used today tend to produce notes with widely differing volume levels.

Acoustic Basses

Acoustic double-basses can be more problematic. The combination of the instrument's relatively quiet acoustic output and the proclivity of most upright bass players to move around means that, other than for classical ensembles, miking it up live is a recipe for disaster. If the bass is fitted with a dedicated pickup system, you will at least be able to get some decent level into the PA, albeit not always sounding much like an acoustic bass. Players will often turn up with a mic wrapped in foam and wedged under the tailpiece. If you're lucky you might be able to EQ this to sound OK, but they are always prone to rattling and also to picking up high levels of spill. Whatever you do, the end result is usually a compromise. Rockabilly-style 'slap' bass players will usually want to emphasise the string slap of an upright bass, so even if their instrument is fitted with its own mic or pickup, you'll need to put a separate mic as close to the strings as possible without it getting in the way of the player,

Conventionally miking an upright bass is fraught with problems, so some kind of contact pickup or instrument-mounted mic is often the only practical approach.

usually around half way up the neck. You'll then need to EQ out all low-frequency content until you just hear a percussive slap. This can be mixed in with the basic sound of the instrument to achieve the desired effect. Ensure that the mic on the strings points away from the drum kit and any nearby back-line amps or monitors to keep spill to a minimum, as the sound coming off the bass strings isn't particularly loud. I prefer to use dedicated acoustic bass pickups whenever possible in live situations.

Drums

A good drum sound always starts with a well-tuned drum kit, and provided that the kit is fitted with heads and snares that are in decent condition, it should be possible to coax any reasonable kit into sounding decent fairly quickly. Worn heads and stretched snares really will make the job much more difficult, and sometimes even impossible.

Where the room acoustics are favourable and spill isn't a problem, a stereo microphone pair a metre or two in front of the kit will actually give a very natural reinforcement of its acoustic sound, but this is never an option for a fully amplified band, as the spill level will be too high. For smaller gigs, a mic on the kick drum may often be enough just to add the required weight to the direct sound heard by the audience. Kick-drum mics are usually dynamic cardioid models able to cope with high SPLs, and often specifically voiced for use with bass drums to give them a distinct advantage over trying to coax an acceptable sound from an inappropriate microphone.

Assuming the kick drum has no front head, or a front head with a largish cut-out, a fairly standard mic positioning would be just inside the drum shell, pointing towards a spot roughly halfway between the shell and where the beater hits. If you don't have a suitable mic-stand, you can sometimes get acceptable results simply by lying the mic on top of the damping blanket that drummers often keep inside their kick drum. A kit with no hole in the front head can be miked in front, if you want a particularly deep sound with no attack, or from behind for more click, as long as you can place the mic somewhere where the drummer won't kick it (or its stand).

Kit miking can be as minimal as just adding some support for the kick, or as full-on as close-miking every drum.

The next step up on the drumkit miking scale is to add a close mic on the snare, or one or two overhead mics. Some live-sound engineers like to use as few mics as possible and, providing the balance is OK with just kick and overheads, are often perfectly happy to dispense with a snare close mic in favour of a more natural overhead-based strategy. Most prefer a snare mic, however, as the snare-driven backbeat is such a fundamental part of rock music. A dynamic cardioid model is usually chosen for this job, generally placed 50mm or so above the snare rim, pointing in towards the centre of the top head. A capacitor mic will give a slightly more open sound with more of the wiry rattle of the snares audible, but you'll probably have to engage the on-board pad to avoid overloads. A second mic is sometimes used below the snare drum to pick up more of the

A dynamic cardioid is usually favoured for snare drum, particularly for a powerful rock drum sound, but you will miss out on some of the subtlety and complexity of the drum unless you are prepared to mike the bottom as well.

wire sound: if you feel you need this, remember that the polarity of the lower mic needs to be inverted using the desk's 'phase' switch, as the bottom mic is facing in the opposite direction to the top snare mic.

Cardioid capacitor or back-electret mics are usually chosen for overheads because of their extended high-frequency response. These should be mounted around two to three feet above the cymbals, at a point that achieves an even coverage of all the drums and cymbals. If you have put close mics on most of the other drums, you can use a significant amount of low-cut filtering on the overheads. In most cases, you'll find that you've already got more hi-hat than you'd like coming down the snare mic and overheads, so you'd probably only resort to a close mic for the hi-hat in a very large venue with a drummer who didn't hit the hi-hats very hard.

Toms can be miked from just above the rim, pointing across to the centre—a key consideration, just as with the snare mic, is that the drummer should be able to play normally without risk of hitting the mics. Dynamic cardioids are a favourite, once again, but there will often be an excessive amount of boom in a close-miked tom, benefiting from a degree of damping in the form of a small amount of tape or gel on the top head.

Drum damping can be achieved by taping folded tissue or cloth to the edge of the top head of the drum, or via the sticky Moongel polymer patches that many drummers have adopted. Damping pads usually go fairly close to the edge of the head, in a place where they won't be hit and where they are not directly beneath a microphone. Take care not to overdamp the drums whilst listening to them on their own, as they'll probably sound dull and lifeless as soon as the other members of the band all join in.

To increase the amount of damping on a kick drum using a 'blanket' damper, push the blanket so that it contacts more of the rear head but, again, don't make it too dead or the sound will lose its punch. While it is always better to get the sound right at source, noise gates can be used to tighten up close-miked drum sounds by setting a short release time.

For bands playing in smaller venues, clip-on drum mics make a lot of sense as they remove the clutter of a separate boom stand for every mic on the kit. The clips, which clamp onto the drum rims, are designed to offer a degree of vibration isolation and make it easier to place the mics where they won't be in the drummer's way. Even the more affordable, all-in-one, drum-mic sets can produce very acceptable results, although the kick drum mic in a budget set often leaves something to be desired,

Be careful not to over-do damping on drums—what sounds right in isolation, may not work in a full-band mix.

so you might want to think about upgrading to a reputable dedicated kick mic.

Acoustic Instruments

Unless you are working with an all-acoustic act, it can be a nightmare getting anything like enough level from a miked acoustic guitar or violin to allow it to compete with amplified instruments. However, where miking an acoustic guitar is practical, a cardioid capacitor mic placed around ten inches away and aimed where the neck joins the body will reliably produce a balanced, natural sound. Avoid aiming the mic at the sound hole as this will just pick up the boomy main-body resonance. For violin, the mic ideally needs to be two feet or so above the soundboard, which makes spill a real problem in an electric band. Even where you can get an acceptable sound by moving the mic closer, the player's movements are likely to cause problems because the closer the mic, the more the level and tone will change. A clamp system that fixes a miniature mic to the body of the instrument can help, but the mic position may have to be less than optimal.

By far the safest bet for acoustic guitar or violin in an amplified band context is to use an instrument fitted with a pickup system, plugged either into an acoustic guitar combo (most of which have balanced DI outputs) or directly into the PA via a DI box. This at least reduces spill to a negligible level, and significantly reduces the risk of acoustic feedback.

Piano

Small venues often won't have an acoustic piano at all, or they'll have an untuned and poorly maintained one, so in most cases, an electronic piano will be a much safer option. If you have to mic an upright acoustic piano, remove the upper covers and hang a couple of mics over the open top of the instrument, perhaps two to three feet apart and 10 to 12 inches above the top of the case. The extended frequency response and sensitivity of capacitor models will give the best results, but you will need to experiment with placement to achieve an even coverage across the strings.

Grand pianos can be miked by opening the lid and hanging a couple of mics about 12 inches back from the hammers and about the same distance above the strings. Again, the exact placement will need to be adjusted to give an even volume across the keyboard. Of course, this doesn't begin to capture a complete picture of the instrument, but close miking is really the only practical strategy in live sound, other than for a solo piano performance, simply because of the amount of spill you have to deal with.

Brass And Wind Instruments

Flutes and whistles tend to need the mic fairly close to the mouthpiece (but not so close that the player is blowing into the mic) to pick up enough level, even though sound also comes out of the end the instrument and any uncovered finger holes. The saxophone is similar, but less problematic as it generates a lot of level, although for pop and rock work the sound of miking the bell is sometimes preferred to the more accurate capture achieved by placing the mic above the bell, aimed around half way up the body. For the trumpet, trombone and all similar instruments, level isn't a problem, so a mic placed 10 to 12 inches in front of the bell will do the job. It is, however, essential to choose a mic that can handle the required SPL, as brass instruments can be very loud—a mic capable of handling 135dB SPL or more is desirable in this application. High quality dynamic models such as the Electrovoice RE20 and RE320 are particularly good for brass, although any general-purpose dynamic instrument mic without a too prominent presence peak should be fine.

Rotary Speaker Cabinet

Used mainly on organ but occasionally on guitar, the Leslie cabinet is the most famous of all rotary cabinets. A crossover splits the audio between a rotating deflector drum placed over the bass speaker and a spinning horn that 'sprays' the high frequencies around like some kind of mad lighthouse. Obviously, both the top and bottom of the cab need to be miked to capture a full-range signal, and the spatial effect is much greater if you are able to mic the top rotors in stereo, assuming you are using a stereo front-of-house mix (a single mic on the bass rotor is fine). The stereo effect will vary depending on the mic spacing and distance, but, as usual, the potential for spill usually dictates a placement of six to 10 inches away from the cabinet louvres. Dynamic mics are often preferred.

'Unusual' Instruments

Faced with unusual instruments on stage, I simply listen to the instrument to see which part generates most of the sound and which way it projects. For example, with an acoustic guitar, relatively little comes directly from the strings: the majority of the sound comes from the body, projecting forwards from the top.

Once I've figured out where the sound comes from I use the length of the sound-producing part of the instrument (in this case the body length) as the initial mic distance. That approach invariably captures a fairly accurate representation of what's going on, but on stage you may often need to get in tighter to reduce spill, especially if the instrument is not very loud. With an instrument that is especially small, you may also need to increase the mic distance or adjust the mic position to leave room for the player's hands.

The basic frequency spectrum of the instrument is the major consideration for mic choice: if it generates a lot of high-frequency detail then a capacitor mic will give the best results, and if there's not much low end you should be able to switch in the low-cut filter as well, to help keep the sound clean. An instrument with more midrange than top-end will usually be fine with a dynamic, unless you need the higher sensitivity of a capacitor model. A bassy instrument will need a mic with a full bass response i.e. not a vocal mic with shaved-off lows, and loud instruments need mics with appropriate SPL handling. Once you learn to ask yourself these basic questions, the mic choice and placement usually become self-evident.

Pre-Flight' Checks

Once all the mics and DI's are in place on stage, you'll need to check each one is working and connected to the right mixer channel. Gently scratching the mic grilles is kinder, both to the mic and any bystanders, than endlessly yelling 'one-two' or, worse, blowing into the mic! Only once you have verified that everything is working should you bother to 'dress' all the mic cables and tape them down. Then you can get on with ringing out the system to identify and address potential feedback issues, before getting into the business of a full sound check.

Working With EQ

In live sound, instrument tonality is often compromised through close mic placement in the interests of minimising spill, so some degree of corrective EQ is frequently required. As the human ear is much more sensitive to narrow-spectrum EQ boosting than to cutting, the most natural sounding EQ results can usually be achieved by cutting the frequencies you can hear too much of, rather than by boosting the ones you feel are lacking. You can demonstrate this very easily by setting maximum boost in a parametric EQ stage, with a high Q value, and sweeping the frequency across the audio spectrum. You'll hear the frequencies being emphasised in a very obvious way, almost like a wah wah pedal being rocked. But if you do the same exercise with a sharp EQ cut, the effect will be far less noticeable.

For acoustic instruments, it is always best to use as little EQ as possible, and appropriate subtractive, EQ combined with broad, gentle boosts where necessary will keep the result sounding as natural and uncoloured as possible. Electric instruments and close-miked drums have greater tolerance to more assertive EQ settings.

A simple 'bracketing' process can help keep things sounding clean, using low- and high-cut filters to attenuate any highs and lows that are not a necessary part of the instrument's spectrum. A filter slope of 12 to 18dB/octave for the low-cut filter is the usual choice, and can usefully be applied to most 'non-bass' sources. Bracketing helps reduce the spectral overlap between instruments that might otherwise tend to conflict, a typical example being synthesiser pad sounds, which often take up a lot of spectral 'real estate' for no good reason—thinning out the lows avoids conflict with bass instruments and kick drums as well as avoiding low-mid boxy build-up, while pulling down the highs prevents the pad from masking cymbals, the upper edge of vocals, and guitars. It won't sound as impressive on its own, but it will still do its job in the mix without masking other instruments.

A parametric equaliser set to a wide bandwidth (low Q) can be used to add subtle emphasis to a particular area of the frequency spectrum, for example, upper-mid clarity to vocals or to warm up basses and kick drums. Shelving EQ is a benign-sounding way of adjusting the general brightness of a sound at the high end, or depth and warmth at the low end, but beware of lifting unwanted noise at either extreme of the spectrum.

A gentle top-cut is actually a good way of moving things backwards in the mix so that other things can be more forward, without having to make them louder.

How To 'Tune' Sweep EQ

The simplest way to locate the frequency range that needs adjustment using a parametric sweep EQ is to set 10 to 15dB of boost (and a high Q setting, if possible), and then sweep it through its frequency range, noting which areas make a positive contribution to the sound and which ones sound unpleasant. If the parametric EQ sections have fixed Q, the results might not be quite so obvious, but significant problem areas will still stand out. For example, a vocal might sound boxy in the lower mid-range (200 to 400Hz), nasal around 1kHz, harsh at 3 to 4kHz,

If you set a midrange filter with 10 to 15dB of boost and sweep it through its frequency range, areas that make a positive contribution to the sound, and those that don't, will be readily apparent.

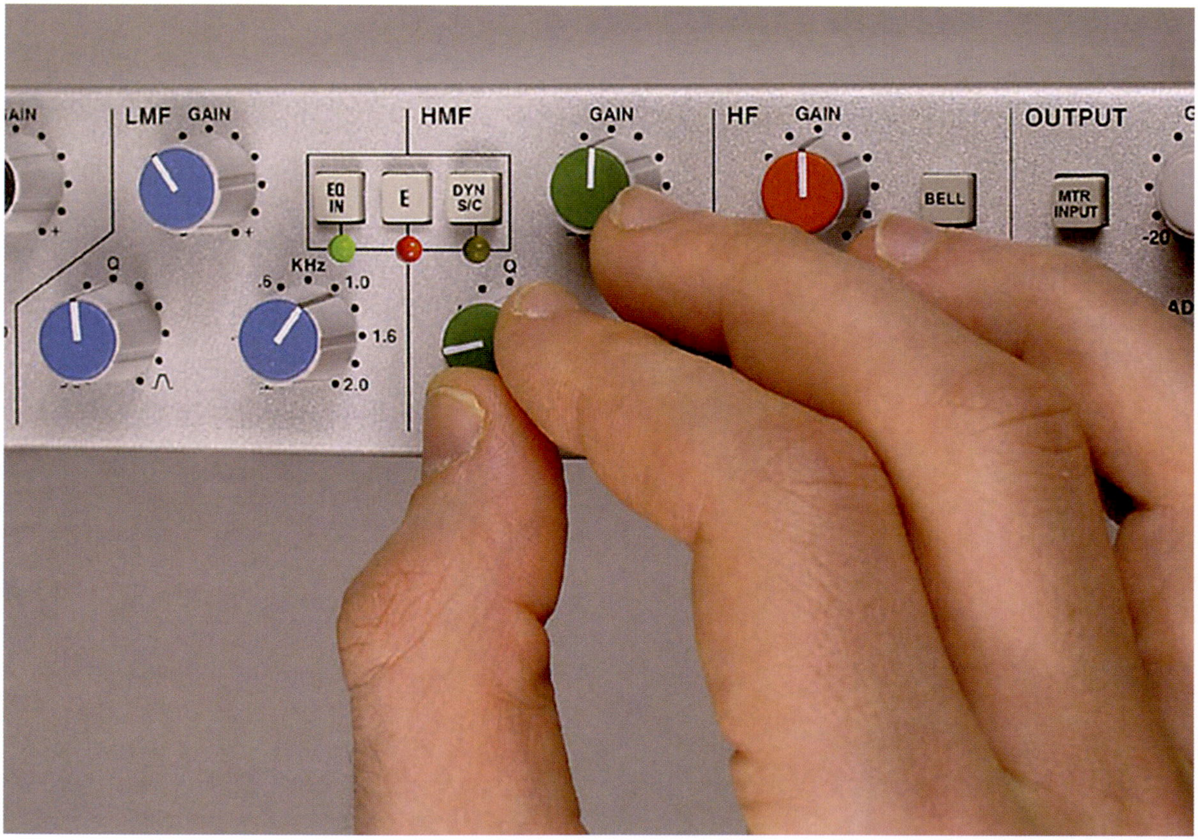

airy above 8kHz, and so on. When you do this with mic signals, you will have to turn down the mixer's main output level, as the narrow-band boost will almost certainly push the system into feedback.

Having identified the ranges that need cutting or boosting, you can adjust the EQ for more appropriate gain and Q values—the high-boost, high-Q setting was simply to find the areas that need attention. If you are already addressing low-frequency problems using a low-cut filter, you should leave this switched in before adjusting any parametric mid controls.

Always check the final EQ'd sound against the EQ's flat or bypass setting, as it is easy to lose focus and apply too much. Equally, don't be mislead by the effects of a simple overall level increase due to overall EQ boosting, as the human ear is easily fooled into thinking louder is automatically better. Some instruments need to sit behind others to create a viable live-sound balance and if everything is EQ'd to sound bright, with lots of presence, the overall result will be extremely fatiguing and difficult to listen to.

EQ'ing Drums

Kick drums may have a fundamental pitch as low as 40Hz, although the bulk of their sonic punch is focussed around 70 to 100Hz. When using speaker systems that aren't designed to reproduce high levels of very low frequencies you can achieve more headroom by trimming off the real low-end, below 50Hz, and although some tonal weight will be lost, the sound will still be adequately punchy in the mix.

The beater 'hit' of a kick drum occurs in the 3 to 5kHz region, and a little narrow-band boost in this area can add 'click' and definition. Too much in the low mids, say 150 to 300Hz, can make a kick drum sound boxy, and will invariably benefit from some low-mid cut to clean up and focus the sound. So to add punch to a kick drum, try boosting at 80—90Hz whilst also cutting the 180 to 220Hz region. Dedicated kick drum mics are often 'voiced' to cut the problematic low mids, and therefore require less corrective EQ to get the desired sound.

Tom toms often suffer from excessive low frequencies, with most of their energy focused in the 80 to 200Hz region, and this is often exacerbated by close-miking. In addition, they'll often resonate in sympathy with other drums and the bass guitar, producing an almost continuous low-frequency drone. However, if you overdo the low-cut EQ, they'll have no punch at all when it comes to drum fills. The stick-hit element of the sound can be enhanced by boosting between 4 and 6kHz, and a significant amount of low cut applied to the overheads can help clean up the overall sound, if the toms are close-miked.

Snare drums cover a wider part of the audio spectrum than toms due to the 'snap' and 'rattle' of the wire snares that give the drum its distinct identity. You can add weight, by boosting the 90–150Hz region while cutting the 200 to 300Hz region to counter any boxiness. The snap of the snares is usually evident between 4 and 8kHz, and a high shelving EQ boost will add brightness without colouration. A low-cut filter will reduce the level of low frequency spill from the kick drum. If you have one, a harmonic exciter may be more successful than EQ in creating snap and brightness if you are confronted with a particularly dull-sounding snare drum.

On small stages the overheads may need to be closer to the kit than is ideal, and their level in the mix reduced to avoid spill from other backline elements. Rolling off some of the lows can help make the kick and tom sound more solid, because it avoids phase-cancellation due to the close and overhead mics picking up the same sound at slightly different times. You still need the overheads to pick out the cymbals and to add stick definition to the drum sound though, so use either a broad boost centred around 6kHz, or use a high shelf EQ to add a little boost at 6 to 8kHz.

Vocal EQ

Having selected a mic that complements the singer's voice, you will still probably need to use a low-cut filter to compensate for proximity effect bass-boost, and also to combat popping. If this starts to have an audible effect on the overall tonality, a little broad-bandwidth boost at 150Hz will help warm up the sound again. Boominess or boxiness generally occurs in the 250Hz region, while the 'nasal' component will be around 1kHz. Harshness can usually be traced to the 2.5 to 3.5kHz region, and sibilance on 'S' and 'T' sounds comes in around 6kHz. To add airiness without harshness, try a little shelving boost at 8 to 12kHz.

Bass Guitar EQ

While a DI'd bass guitar produces frequencies all the way up the audio spectrum, the amplified sound tends to roll off fairly steeply above 4 or 5kHz because of the limited response of the speaker cabinet. Boosting between 50Hz and 80Hz will lift the low bass provided that the PA system can stand it, but it is best to filter out anything below this range to reduce the headroom-eating subsonics. The 200 to 300Hz range is a crucial area for electric bass guitar, as this is where the character of the sound is defined and where the actual pitch of the bass notes is most easily discernable. A subtle EQ boost in this range will help the bass line stand out more in a busy mix. If finger noise and fret rattle are a problem, gentle cut above 6kHz will generally be enough to stop them being too obtrusive.

Synthesised basses can always be adjusted on the instrument itself, but shaving off the ultra-lows is still a good idea as electronic instruments can generate very high levels of almost inaudibly low frequencies that would otherwise just stress your speakers and consume amplifier headroom.

EQ'ing Acoustic Guitar

Acoustic Guitar EQ depends very much on the musical context. For example, a solo acoustic guitar usually needs to sound fairly natural, whereas the same instrument playing a supporting role in a pop mix will almost certainly work better with the low end trimmed right back. This will invariably sound too thin when heard in isolation, but it's what works for the mix that matters. You can also add a little 'air' to the top end using a shelving EQ boosting around 8kHz. Some pickup systems tend to over-emphasise the body resonance of the instrument, so use the 'sweep and tune' approach to locate this frequency and cut as necessary.

Electric Guitar

Electric guitars reside in the 80Hz to 4kHz range, so you can usually switch in your 80 or 100Hz low-cut filter without detracting from the weight of the sound at all. Amp hiss can be reduced by trimming the highs above 6kHz, and muddiness tackled by cutting in the 150 to 300Hz region. The characteristic tonal bite usually resides in the 2.5 to 5kHz range. Some modern rock guitar sounds pair ultra-saturated distortion with a strong midrange 'scoop', most of which should come from the amplifier settings, although the mixer's EQ can be used to lend it a helping hand if the midrange is still too thick-sounding.

Problem Noises

Mains hum, which is evident with many valve guitar amplifiers, occurs at 50Hz in the UK, Europe and many other regions (60Hz in the US and some other parts of the world). However, harmonics above the fundamental frequency are often also audible, so it's worth cutting at 100Hz, 150Hz, 200Hz and so on, using the narrowest bandwidth setting that does the job. Most mixer EQs don't have enough EQ bands to do this effectively, so in practice, you'll usually have to settle for cutting the fundamental and perhaps the first harmonic. Some lighting dimmers can generate a very broad-spectrum buzz, which sounds much more aggressive than hum and which can't be removed using EQ. This can affect a number of sources, but is particularly prevalent with electric guitars using single-coil pickups.

Graphic EQs are popular for adjusting systems to compensate for room acoustics, or 'ringing-out' problem feedback frequencies.

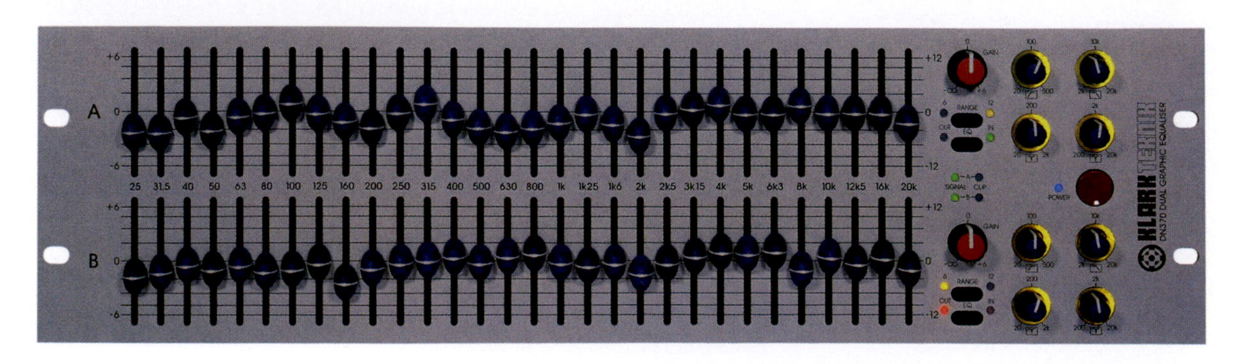

THE 'SMILE' CURVE

You'll often see graphic equalisers set up in what is called a 'smile curve,' especially when playing back pre-recorded music. The human hearing mechanism actually perceives a different frequency balance at different sound levels—at higher SPLs, we hear more low-end and more top-end, relative to the midrange. The 'loudness button' on many consumer audio systems is a preset EQ that compensates for the loss of bass and top-end when listening at very low levels. In the context of live sound mixing, pulling down the midrange slightly, or boosting the highs and lows, can make a mix sound clearer and more punchy. If your mixer allows sub-grouping you can use this idea to help keep the vocals audible by introducing a gentle midrange scoop to the back line mix, but not to the vocals.

PANNING—KEEP IT TIGHT

When mixing a show in a small venue, it pays to work in mono or keep any stereo panning fairly narrow, otherwise you end up in the situation where audience members closer to one of the speakers than the other hears mainly one side of the 'stereo', which might, for example, mean hearing mainly just the high notes or low notes from a widely-panned piano. The same applies to panning or ping-pong delays. However, stereo reverb can be left at full width.

It pays to be fairly conservative with panning when running a stereo front-of-house mix as a good proportion of your audience will not be hearing both sides of the system equally.

Chapter Twelve
The Role Of The Mix Engineer

So far we've looked at how to choose and rig the key equipment required to put on a live-music event, but the best equipment in the world is of no use if the person operating it doesn't know what they are doing. Small and medium-sized venues may seem on a different scale to what you see at major tours and stadium gigs, but that doesn't actually make the mix engineer's job any easier, or less important, and as we'll see, it can, in some ways, be more difficult. It's time to take a closer look at that most vital of sound-system components, the 'sound guy'.

In days gone by the 'sound guy' was always a friend of the band who happened to own a van, but things are a little more sophisticated today, and audiences have come to expect a certain level of sound quality and professionalism, even just from a band at their local pub. Ideally, every band would have their own mix engineer, who is familiar with the material and every band member's monitoring preferences. However, the combination of a decent technical background and a degree of musical empathy, never mind the level of commitment required, is not that easy to come by.

Many bands—possibly even most bands on the small gig circuit—will end up having their performance mixed by someone who has never heard their material before, and certainly as a PA provider, I'm often asked to work with bands I've never met before the night of the gig. When working with an unfamiliar band, I find it helps to check out any on-line songs and videos

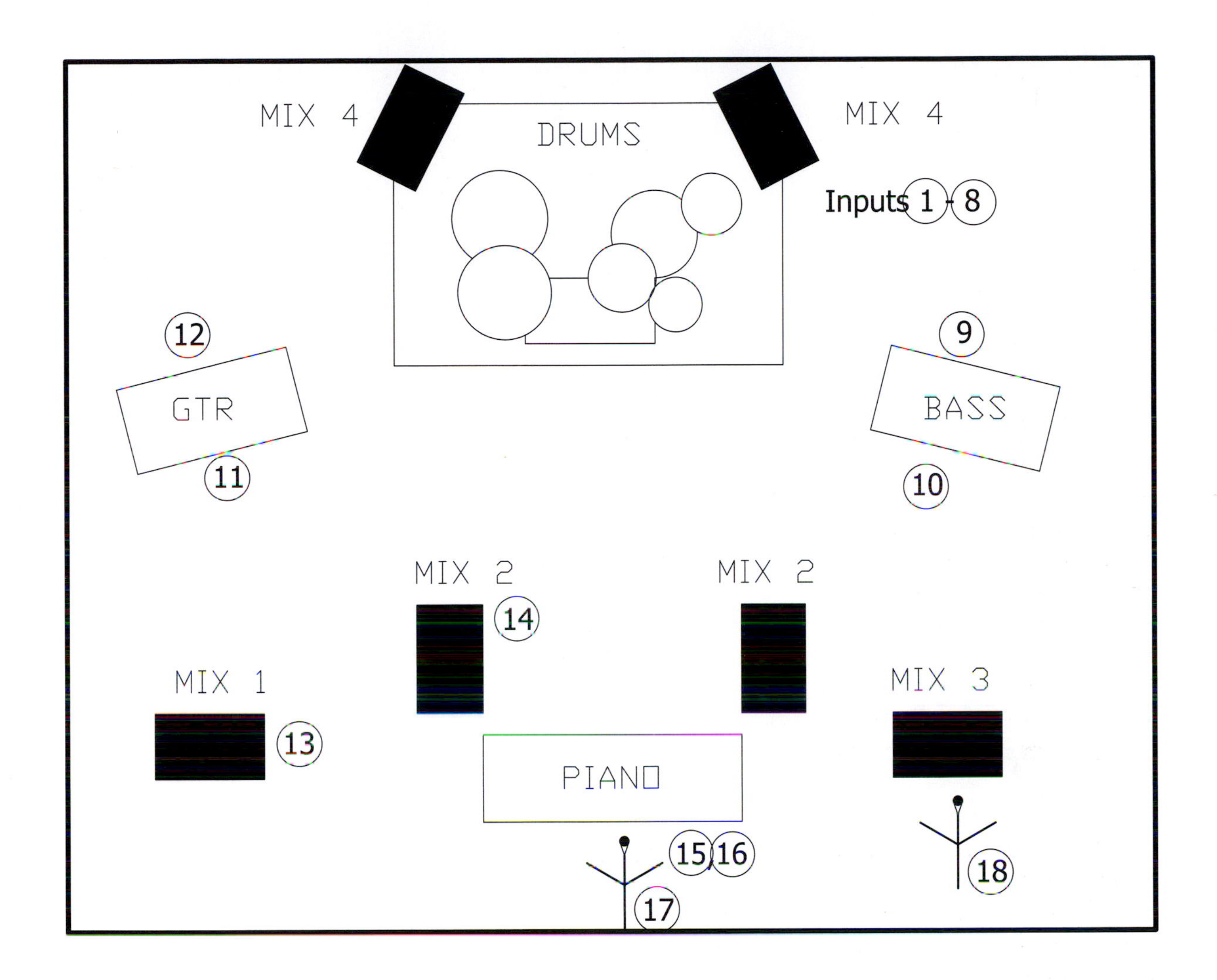

A stage plot sent well in advance certainly aids your chances of getting your PA requirements met at an unfamiliar venue.

they may have, as that will give some idea of how they want to sound and setup on stage. It's always a good idea to get in contact before the gig, too, to ask if they have any 'special requirements', so you won't be surprised by anything on the night. At the very least you need to know how many instrument and vocal mics they need, as well as the number of DI boxes.

For larger gigs, bands with management will often supply a 'system specification', detailing their requirements. In my experience, however, these all too often specify a front-of-house rig that would be overkill for the venue, and an impractical number of separate monitor mixes.

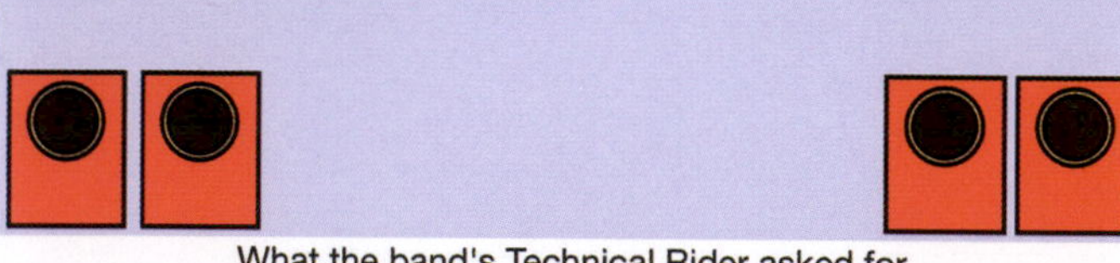

What the band's Technical Rider asked for...

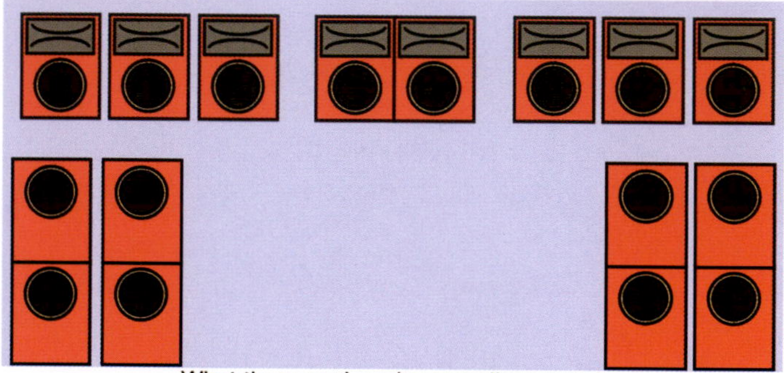

What the sound engineer really wanted...

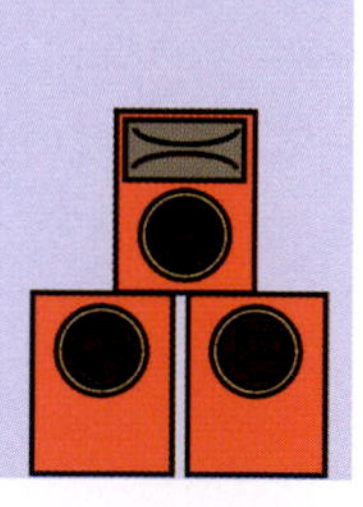

What the accountant hoped for...

What the venue actually needed.

So true . . .

There's nothing worse than being on the wrong end of a badly-operated PA system, as I've discovered more than once when playing with my own band. On one occasion I asked if a monitor could be turned down as it was deafening me, only to be told by a member of the audience that the sound desk was unmanned and the 'engineer' had gone to the bar! You wouldn't expect the guitar player to walk off stage and go to the bar half way through a set, so why should a mix engineer think that was OK?

If you engineer live sound for a band on a regular basis, I believe you should be just as committed to the performance as the musicians, and that means 100% concentration from the start of the sound check to the time when the PA is powered down at the end of the gig. You should know the material almost as well as they do: who sings which songs, who takes solos and when, and which effects are required and when. If you need a cheat-sheet to help you remember, that's fine—just make sure you don't lose it.

If you find yourself mixing for an unfamiliar band, the analogy for your role would be far more like a musician at a jam session, where you'd have your ears and eyes wide open, trying to respond both to what is happening and to anticipate what is likely to happen next, being ready to ride the faders to make sure solos and lead lines come across properly, and so on.

Preparation

In reality, the role of any sound person with a professional outlook starts well before the performance, beginning with the planning of what's needed for the gig. Depending on the type of PA gear you have, you may be able to scale the system to the venue or the type of act, so knowing the size and layout of the space you'll be working in is always a distinct advantage. You should also have a contingency plan in case some key piece of equipment breaks down. You'll earn extra points from the band if you can get them out of a fix of their own making, too.

Try to imagine what could go wrong and then figure out what else you need to take with you to get yourself out of trouble.

Even if you can no longer amplify the backline, you always need to have a backup strategy to provide vocal amplification—so long as you have that, the gig can be completed in some form.

It seems to be considered the sound person's responsibility to be able to provide sundry small items that might be needed during the gig, such as spare instrument cables, mains extension cables and distribution boards for the backline, spare batteries for pedals and radio mics, and of course plenty of gaffa tape. It doesn't hurt to keep a pack of guitar strings or two

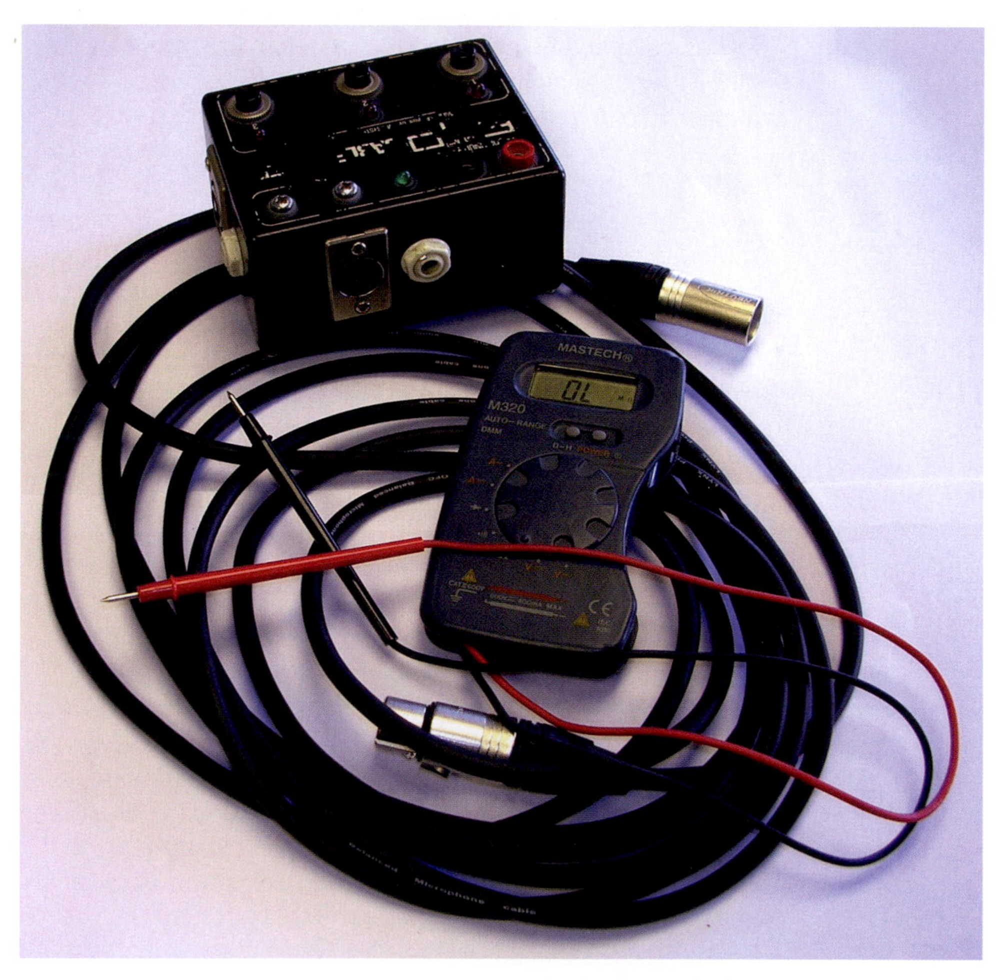

A few simple tools that allow you to check for continuity, such as a cable checker and a multimeter, can save you an awful lot of troubleshooting time.

in your bag, too, just in case. That's in addition to the PA essentials such as spare mics and stand clips, XLR cables, and an emergency tool kit and torch for running repairs! Complete 'roadie' tools sets are available that cover most eventualities, including suitable wrenches for adjusting guitar bridges or locking nuts. But even if you are a band that doesn't have an engineer, mixing yourselves from the stage, it helps if one individual takes responsibility for ensuring that the 'spares' requirements are covered.

I can't stress too highly that mains cables should be inspected regularly and appropriately tested, and using your own mains socket tester and RCD circuit breaker provides an extra layer of safety against electric shocks.

Setting Up

Always try to get to the venue in plenty of time to load in your gear and get it set up before the audience comes in. Diplomacy and a willingness to compromise goes a long way in dealing with the people running the venue, as there's often a yawning chasm between where they say it's OK to put your PA speakers and where the laws of physics suggest they should be placed! The same applies to the mixing desk—we all know where we'd like it to be situated, but in reality you'll often find yourself jammed up against a side wall or even stuck on the side of the stage . . . if there even is a stage, that is!

Running alongside such territorial negotiations is the need to take health and safety considerations very seriously, and if you operate a PA system for people other than your own band, you should make sure you have third-party insurance. Cables are a potential trip hazard, and while band members may be used to wading in a sea of cables, audience members need to be protected from any cables encroaching into public areas, such as wires connecting the PA speakers, or a multicore running from the stage to the mixing position.

Running cables around the room edges or over doorways, and taping them down, may often be all that's required, but rubber matting taped down over cables is also very useful in some situations. Occasionally you might come across a venue owner who has a 'no gaffa tape' rule because of the sticky residue it can leave behind, so you may have to organise your cables to follow a different route—in which case spare cables to use as extenders are useful.

Tablet-controllable digital mixers can be very attractive when you're faced with difficult venues, as the mixer itself can be left on stage along with all the attendant cabling. The mix engineer then only has to find a suitable seat where they can see and hear what the audience sees and hears, to pick up on any body language or gestures indicating a problem. A further benefit of these systems is that you can carry the tablet around to check and adjust the sound from various points in the venue.

Visual Cues

Mixing a band for the first time, just watching the performers closely will give you a fair idea of what is about to happen, so placing the mixer with an uninterrupted sight line is important.

Line of sight is particularly important when you're working with unfamiliar musicians—or if your regular band is prone to improvisational departures from their normal arrangements—as the visual cues provided by the musicians' body language are extremely important in anticipating what is going to happen next. Watching the singer during a performance will often give away the fact that a guitar solo is about to start, as they back off from the mic and/or glance at the player about to take over. The same is true of guitar players who usually either make a move to their pedalboard or adjust the volume control on the guitar—it doesn't take long to learn the visual cues of any typical group of musicians. Having some basic musical knowledge helps too, as the song structure itself will provide clues as to where changes are likely to occur.

Balance

The purpose of putting all the instruments through the PA system is not just to make everything louder in large venues, but also to give the engineer control over the band's balance. In small venues, however, if the backline is turned up too far, it can be impossible for the engineer to achieve a proper balance. A guitar player or drummer that is already too loud for the venue deprives you of all control and you may have to be very assertive to get them to adjust their volume.

Snare drums are always the most overbearing part of the kit in smaller spaces, so it's a great idea for drummers to carry a quieter, backup snare drum—this will usually be more effective than trying to damp down a loud one, as this usually compromises both the tone and playing feel. There's also the option of using quieter drumsticks, such as ProMark Cool Rods or Lidwish sticks, although not all drummers get on with these as their playing feel is a little different from normal sticks.

I know that drummers are often recommended to use hearing protection, but in my experience this invariably leads to them playing even louder. If we all have to risk our hearing by standing close to them, the least they can do is share the risk and leave out the earplugs! Many musicians find it impossible to play or to pitch vocals properly when wearing ear plugs or attenuators, so there may be no practical defence against an over-loud drummer with 20dB attenuators in his ears!

The more radical solution is to use an entirely electronic drum kit, something that's becoming ever more popular for small-venue performance. Their big advantage relative to a real kit is that their on-stage volume is controlled entirely by the amplifier into which they are plugged. The more sophisticated kits featuring mesh heads, have a good stick response, and can deliver a 'produced' studio sound on stage. But even the more affordable kits can feel very natural to play and sound OK too, especially if used with real cymbals. I feel there's also a future for acoustic kits augmented by a fully electronic snare drum, as not only can the volume be adjusted, but the sympathetic rattling suffered by acoustic snare drums is avoided. Relatively compact full-range monitors are available specifically for use with electronic drum kits.

Perhaps the biggest attraction of electronic drum kits is that their volume is totally controllable, allowing the drummer to play normally whilst the sound level is entirely controlled by the needs of the front-of-house and monitor mixes.

Big guitar stacks may look impressive, but a band using modestly-powered back-line amplification fed via the PA will invariably sound better. As I've pointed out already, for typical pub and small club work, guitar combo amplifiers rated at around 20 to 50 watts and bass/keyboard amplifiers rated at around 100 watts will invariably be more than adequate.

While bass and keyboard stage amps can be turned up or down without significantly changing the sound, guitar players always know exactly where the 'sweet spot' of their amplifier is, and in smaller venues it is often 20dB higher than the front-of-house engineer would like it to be! Managing conflicts in the area of on-stage volume calls for a degree of diplomacy, and if you can't persuade the player to use a smaller amp, assuming one is even available, then you have to try things like pointing the amp in a different direction (into the side or rear of the stage for example), or perhaps even making more use of pedals to get the sound, so that raw amplifier volume is less of an issue. Try to set up the amp in a position where it sounds adequately loud to the person playing through it, which may mean putting it on a stand rather than on the floor, as the louder it sounds to them, the less likely they are to want to turn it up.

Artificially manipulating the monitor mix can sometimes help too in taming an over-loud player. For example, you can turn down their overall monitoring level slightly while increasing the level of their instrument in the monitor mix so that they have to play less loudly in order to hear the rest of the band.

How far back into the stage area the amps are placed is also a significant consideration, as the drummer will only hear a boxy, coloured sound from the back of sealed-back cabinets if they are placed anywhere in front of the kit. Open-back combos, are rather better in this respect. This is all the more important where the amount of stage monitoring is limited, with perhaps only a vocal monitor. The better the players can hear each other, the better the performance you'll get out of them, and how the backline is positioned on stage can make a significant contribution. Angling the guitar cabinets so the drummer can hear them more clearly is one simple option.

Pearl

Monitor positioning is often dictated by the size of the stage, but you can often make them more audible to the frontline performers by pulling them back from the edge a little.

Stage monitors need to be placed both where they can be heard best and where they are least likely to provoke feedback. Vocal mics need to be set up taking the monitor positions into consideration—if you want to be really accurate about this you could make up a template that can be held over the mic to indicate the direction of the mic's dead zone(s). You also need to be aware of loud backline sources directly facing into vocal mics and adjust their position on stage, if possible.

Sound Check

Once the PA system has been completely set up and tested, we come to the business of setting a basic level for the sound check. In smaller venues, I find the best way to get a good initial mix is to first establish how loud you can get the vocals, leaving an appropriate safety margin, and then try to balance the rest of the band to that reference. I also usually have a friendly chat to vocalists about where not to point the mic and about not obscuring the lower part of the basket with their hands, as either can aggravate feedback problems.

Balancing the band first and then expecting the vocals to compete just doesn't work in smaller venues, as the feedback threshold is often rather lower than might be considered ideal. Automatic feedback suppressors can help to a certain extent, clawing back some level before feedback again becomes a problem, but they offer only a limited improvement and some of the cheaper ones can affect the tonality of the sound. As the audience arrives (assuming they do!) you may be able to perceive a damping down of the acoustics, which should improve your feedback margin slightly.

Unless you have a tablet-controlled digital mixer that allows you to make monitor adjustments when standing on stage listening for yourself, you'll have to rely on the band members to tell you when the monitor mix is right for them—all too often they'll want everything louder than feedback will allow. Checking monitor mixes using headphones or a wedge monitor close to the mixer is of limited use in smaller venues as the direct sound from the band tends to obscure what's happening in the monitor mix, so you may need to go onto the stage when the band is sound checking and listen to the monitors yourself.

Make sure all the major channels are marked up clearly on the mixer—you don't want to be pulling things down at random trying to locate the one that is feeding back.

The mixer channels should always all be correctly marked up on a strip of masking tape or white electrical tape—however much you think you know where everything is, you may not be able to instantly find the right channel if you need to mute something in a hurry.

The better your preparation, the easier the job, so what could possibly go wrong? Of course, you'll probably need to keep your eyes open for drooping boom stands, creeping kick drums, flying drumsticks, and kick drum pedals that have somehow become detached from the rim . . . Actually, it's a rare gig when you don't have to dash to the stage at least once on a mission of mercy!

What, No Sound Check?

When doing small-scale outdoor gigs with several bands playing one after the other, it's highly likely that you'll only be able to soundcheck the first band. My approach is to leave plenty of headroom on all the channels and then try to fine tune the

▲
At small-scale outdoor gigs, it is highly likely that none of the acts will get a sound check, but it shouldn't take more than a few bars for a reasonably adept engineer to get an acceptable balance together.

balance within the first few bars as each new band starts. It helps to make clear notes on which mics are where when changing over between bands, and if you are able to use most of the mics and DI's for the same functions within each band, the amount of adjustment necessary to rebalance the sound shouldn't be too great. For this type of gig, I also find it helps to run two rows of masking tape across the mixer: one left in place to identify which mic or DI box feeds which channel and another blank strip below it than can be customised for the next act to show which mic is on which sound source.

A very 'quick and dirty' way to establish the feedback threshold is to set the channel faders to their maximum position and then turn up the gain trim, working on one channel at a time, until you hear the first ring of feedback. At that point, pull the channel fader down to unity gain and move onto the next channel following the same procedure. Once set you should have a few dBs of headroom before feedback on each mic channel, and in addition I usually drop the master fader by a further 5dB or so to give me a bit more headroom.

Mixing From The Stage

Mixing your own gig from on-stage is quite a challenge, but once you've set an initial balance, using a long guitar lead or radio mic system to see how things sound out front, what it really comes down to is managing any changes that need to be made between songs. For example, do the effects need to change or does an instrument need to come up in level for a solo? Having said that, the mixer still needs to be close at hand in case feedback occurs so buying a suitable stand is a must, as you can't rely on the venue having a suitable table of the right

Remote controllable mixers, using tablet computers and smartphones, have made mixing from the stage less of a compromise.

size and height. If you have a remote-controllable digital mixer, a purpose built tablet-computer support that fits onto your mic stand is ideal as you don't then need to be close to the actual mixer.

The first and most basic requirement when running your PA from the stage should be a footswitch to turn the vocal effects on or off—there's nothing worse than hearing reverb or delay on the mics when talking to the audience between songs. Ideally this should have an LED indicator so that you can see at a glance whether the effects are active or not.

If you are using personal vocal processors that connect between the mic and the PA, each vocalist can be responsible for selecting and muting their own effects when necessary. Models are available that can be fixed to a mic stand, keeping all the controls very accessible, and some are even operated

Personal vocal processors give some of the control of the vocal sound back to the performers.

via a touch screen, so there's no excuse for not taking control into your own hands.

Guitar Solo Boosting

Guitar players manage the volume changes required for solos in different ways. A channel-switching amplifier allows selection of clean and dirty sounds, but what if the solo requires only a boost in level rather than a significant change in sound? Some players use only their guitar's volume control, pushing up the level for the solo and adding just a little more dirt to the sound without radically changing it. But where the amount of overdrive needs to remain exactly the same, with only the level changing, there are simple, affordable units that plug into the effects send/return loop, to switch in a precise amount of attenuation—you set your sound up with the attenuator on, and switch it off to achieve a solo boost with no change in tone.

Of course, many players rely on overdrive and distortion pedals to shape their sound, and these offer plenty of possibilities to preset the necessary lift for a solo. Dual-channel overdrive pedals allow the user to set up a predictable amount of solo boost while still having full control over the amount of overdrive added. Even though digital multi-effects and modelling amplifiers allow the user to switch between presets that store both the sound and level settings, it seems that many guitar players still prefer to choose their own individual pedals to make up their pedalboards and control their levels that way.

With A Little Help From Your Friends

Even when mixing from the stage, friends and partners of the band can help out—so long as they have a little experience and good ear—by discretely signalling to the band members using prearranged gestures. In my band's case, for some reason this usually involves a lot of pointing at my guitar amp followed by pointing at the floor . . . I'm sure you get the idea. If the vocals just aren't loud enough out front, it helps if somebody you can trust is able to let you know about it before you come off stage at the end of the gig!

And afterwards . . .

At the end of the gig there's the packing away to attend to. Packing the mics away first avoids damage or loss. Connecting all the XLR cables end-to-end and then winding them onto a garden-hose drum prevents them from tangling. I've also got into the habit of fitting velcro cable ties to the ends of jack leads and IEC mains cables to keep them tidy and untangled when coiled. These separate cables then go into a heavy-duty bag along with the DI boxes and other small items. Once all the cables have been dealt with, I collapse the stands and pack them in their bags. Whichever way around you do it, having a regular routine during setup and breakdown generally helps everything go more smoothly, especially when you are rushed or tired.

It helps to have a routine that you stick to for packing up after a gig. If you know where everything goes, you'll stand a better chance of knowing what you haven't got.

If possible, have an assistant or band member present to keep an eye on the equipment during the de-rig as it's not unknown for gear to get mis-packed or stolen during the general chaos of packing away and loading. It's always worth having one more check around the venue for items you may have missed, too and if you really want to endear yourself to the venue owner, remove any evidence of your having been there, such as bits of Gaffa tape, old set lists, hooks, cable ties and so on. Oh, and don't forget to get paid!

Here's a quick recap of the important points.

- Where possible, make sure the main PA speakers are positioned well in front of the band.
- High-quality speakers with a controlled, even dispersion pattern will give the best results, with compact line arrays offering further benefits when working in small venues with low ceilings.
- Mics with strong presence peaks encourage feedback at the frequency of the peak.
- Always try to position your speakers to minimise the amount of reflected sound that gets back into the mics.
- Make sure vocalists sing as close to the mics as possible—the louder the source, the less gain will be needed.
- Ensure the monitors are aimed at the least sensitive areas of the microphones.
- 'Ring out' the system to establish the maximum vocal mic level you can get, leave some safety margin, and then balance the backline to that level.
- Use third-octave graphic equalisers or automatic feedback suppressor to tame feedback hot spots. Automatic feedback killers will gain more level and affect the sound rather less than a graphic EQ, though some tonal changes may still be evident.
- While a definitive cure for acoustic feedback has yet to be achieved, you'll find your situation much improved if you take the time to understand how and why feedback occurs and then take the above steps to minimise it.

Chapter Thirteen The Technical Stuff

You don't need a science degree to create a good live-sound mix, but it is worth taking the time to understand the key concepts and terminology involved. It will certainly help you interface all your gear correctly, and it just might save you from making an expensive mistake!

Decibels And Headroom

Audio 'tech-speak' invariably includes numbers with the letters dB written after them, and these 'decibels' can initially be confusing. The commonly encountered terms dB, dBu, dBm, dBv and dBV all relate in some way to audio operating levels, as explained in Chapter 4. We learned that the decibel or dB is simply a means of expressing the ratio between two signal levels on a logarithmic scale, and that allows us to represent the scale of human loudness perception as a decibel range of, say, 140dB rather than as a linear ratio that would be millions to one.

The meters on analogue gear show 0VU at their nominal 'happy' working level, and that still leaves some headroom to accommodate peaks. In contrast, digital systems show the entire dynamic range right up to 0dB Full Scale where clipping occurs, so you have to create your own headroom by leaving 10 to 20dB of safety margin.

'Line level' signals are output by mixers, rack-mount effects units, tape machines, MP3 players and many electronic keyboards. However, mic-level signals, which are very much smaller, are output by microphones and most DI boxes.

Both mic and line signals should be carried via balanced cables where possible to reduce an audio connection's susceptibility to interference and ground loop hum, and to permit the safe operation of any non-phantom powered mic connected to a mixer input that has phantom power applied to it. While unbalanced line-level cables present few interference problems, provided that they are short, it is always best to use balanced cables wherever the equipment allows. (The concepts of balancing and phantom power were introduced in Chapter 4).

SPL (Sound Pressure Level)

The level of sound that humans can perceive ranges from the dropping of a pin to a gunshot—which is a range of several million to one! We use the decibel again to express Sound Pressure Level or SPL, with 0dB SPL (a level defined as 20 micro-Pascals) considered to be the lowest level of sound that can be perceived. A change in air pressure of one Pascal is equivalent to 94dB SPL, but as we don't have built-in Pascal-o-meters, it might be more useful just to give a few examples of real-life SPLs.

Normal speech in a quiet room falls into the 40 to 60dB SPL range, while roadside traffic noise might be as high as 80 or 90dB SPL, and is comparable to the level you might listen to music. Gig levels, even in small venues, can reach 100 to 110dB, and by the time we get up to around 130dB, the level is physically painful and even a short exposure can result in hearing damage. At 194dB SPL, or just over 101 Pascals, the pressure fluctuations are equivalent to 'one atmosphere', and it isn't possible for the sound to get any louder without being grossly distorted—because the air itself clips as it reaches a perfect vacuum on the rarefactions—as in the case of a supersonic shockwave bang.

Balanced Connections

In an unbalanced connection, such as an ordinary guitar cable, the signal is conveyed on the inner wire , and the cable's screen—which is there to prevent interference reaching the signal wire—is connected at both ends to carry the return audio current and complete the circuit. However, this can give rise to hum problems in complex audio systems, as described in more detail in the section on ground loops below.

In contrast, a balanced cable contains two conductors twisted together to carry the signal, encased in a conductive outer screen, and as the screen plays no part in the audio circuit it isn't always necessary to have the screen connected at both ends, though in most cases it doesn't cause a problem.

The 'hot' and 'cold' signal wires 'see' identical source impedances and feed a receiving device with identical load impedances, so that any electromagnetic interference will induce an equal voltage in each wire. Being equal on both signal wires, they will cancel out in the 'differential' input stage at their destination as a balanced input circuit is designed to sense the voltage difference between the two conductors. If a balanced source is connected to an unbalanced destination, the signal level may be halved (reduced by 6dB), or may remain the same, depending on the design of the balanced output stage driving the cable and the way the cable is wired.

Another commonly-used balancing system, sometimes known as impedance-balanced, works by driving the hot conductor of a balanced line with the signal, while connecting the cold conductor to ground through a resistor that presents the same impedance as the driven output. This arrangement offers the same benefits as any other balanced output, whether from a transformer or active devices, when fed into a balanced input, but has the advantage that the signal level remains the same if it is used to feed an unbalanced input.

The key point to appreciate is that noise currents induced into a balanced cable's audio screen don't directly affect the audio signal, as the screen does not double as the audio return path. That's why in a situation where ground-loop hum occurs it is permissible to disconnect the screen at one end to break the loop. This is often done by means of a ground lift switch on one of the connected devices rather than by modifying the cable. Note, however, that where a phantom powered microphone is connected, the screen must be connected at both ends of the cable to carry the phantom power ground current, which is essential for the system to work.

For balanced microphone signals and some line-connection applications, the three-pin XLR connector offers the most rugged solution, although line-level connections are often made via quarter-inch TRS jacks. Neutrik Combi sockets have a jack socket in the centre of a set of XLR-pin sockets and can accept either type of connector.

Resistance And Ohm's Law

Electrical resistance becomes important once you start to connect different pieces of equipment together, and describes a circuit's opposition to DC (direct current) flowing through it, much as a narrow hosepipe might restrict the flow of water. The higher the resistance of a circuit (the narrower the hosepipe), the more voltage (water pressure) you need to push a given electrical current through it. Conductive materials such as copper and aluminium have a low electrical resistance, so current flows easily in them, whilst materials, such as rubber, bakelite, glass and some types of plastic have very high resistances and are therefore less conductive. These are known as insulators, as a negligible amount of current can pass through them.

Technically, resistance really only applies to DC voltages and currents, and is measured in ohms. The simple formula known as ohms Law that we all learned at school takes the form: $R = V/I$, where R is resistance (in ohms), V is the voltage across the circuit, and I is the current (in amps) flowing through the circuit. If you know any two of the three values, you can use Ohm's law to work out the unknown one.

OHM'S LAW

If you know any two of the three values in the equation, you can use Ohm's Law to work out the remaining one.

$$V = IR$$

Voltage = Current (in amps) times Resistance (in ohms)

$$R = V/I$$

Resistance = Voltage (in volts) divided by Current (in amps)

$$I = V/R$$

Current = Voltage (in volts) divided by Resistance (in ohms)

POWER CALCULATIONS

Electrical power is defined as current multiplied by voltage:

$$P = IV$$

Power (in watts) = Current (in amps) times Voltage (in volts)

And by applying this relationship to Ohm's law we get: Power in watts = V (squared)/R, or alternatively I (squared) x R.

$$P=V^2/R$$

Power in watts = Voltage squared divided by Resistance (in ohms)

or alternatively

$$P =I^2R$$

Power in watts = Current (in amps) squared times Resistance (in ohms).

Impedance

When dealing with alternating (AC) voltages such as audio signals, the circuitry ceases to behave as a pure resistor, and we talk about its 'impedance' (often abbreviated to 'Z') to current flow which may vary depending on the signal frequency. In a purely resistive circuit, resistance and impedance are the same thing, but most electronic circuits also include what we call reactive components—capacitors and inductors —which are frequency dependent. So, impedance can be thought of as 'AC resistance', and although still measured in ohms, it may have very different values at different frequencies.

'Input impedance' relates to how much current the input terminals of a device draw from a signal source at a given frequency. The lower the input impedance, therefore, the more current the sending device will be required to supply. Similarly, 'output impedance' is a measure of how much current an output stage can supply at a given frequency, with a lower output impedance indicating that the unit can deliver more current.

In audio we tend to be concerned mainly with signal voltages rather than impedances, so the usual strategy is for device inputs to exhibit a much higher impedance than device outputs. This ensures that minimal electrical loading takes place when connecting a signal source to a receiving device, and applies whether we are talking about microphones connected to preamps, or mixer outputs into power amplifiers. It is a technique called 'voltage matching' and Line-level equipment usually has an input impedance of several tens of thousands of ohms—47 kilohms being a typical figure. The output impedance, on the other hand, is typically just a few tens of ohms.

The same is true of sources such as microphones—the mic preamp of a mixing console usually has an input impedance of roughly 1.5k kilohms, whereas a dynamic microphone designed to be plugged into it will have an output impedance of 200 Ohms or less.

One reason audio equipment is generally designed with such a low output impedance is that it allows long cables to be driven without significant signal loss. When using cables in excess of

10 metres, low-impedance sources minimise signal degradation caused by the cable's own impedance (a cable is equivalent to a small series resistor and parallel capacitor, with a bit of inductance thrown in!).

One significant exception to the low-impedance source rule, though, is the electric guitar with passive pickups. These have a relatively high output impedance. Consequently, it is important to keep the cable length reasonably short, and to choose a cable designed for guitar use, as excess cable capacitance forms a low-pass filter with the source impedance and can result in treble loss.

Loudspeaker Wiring

Speakers have a nominal impedance, specified in ohms, which is essentially their DC resistance, but you have to be aware that this changes with frequency, and amplifiers have to be designed and specified to take account of that. When connecting speakers in series, the positive terminal of one speaker connects to the negative terminal of the next one and so on. Where multiple speakers are daisy-chained in series, the formula for working out the total impedance is simply: $R(\text{total}) = R1 + R2 + R3 \ldots$

COMBINING SPEAKERS IN SERIES

$$R(\text{total}) = R1 + R2 + R3 \ldots$$

The total impedance is the sum of the nominal impedance of all the speakers connected in the series chain.

Parallel connection resistances or impedances can be calculated using the formula: $1/R(\text{total}) = 1/R1 + 1/R2 + 1/R3$ etc. It is sensible when connecting multiple loudspeakers to use models of the same impedance so that the power is shared equally between them. When hooking up speakers in parallel, all the positive terminals should be connected together, and all the negative terminals connected together.

The individual drivers within a cabinet may be wired together in series, parallel, or a combination of the two, in order to generate the requisite combined impedance for the cabinet as a whole.

COMBINING SPEAKERS IN PARALLEL

$$1/R(\text{total}) = 1/R1 + 1/R2 + 1/R3 \ldots$$

The total impedance is the reciprocal of the sum of the reciprocals of the nominal impedance of all the speakers connected in the series chain (the reciprocal of a number is simply 1 divided by that number).

➤
Series and parallel connection.

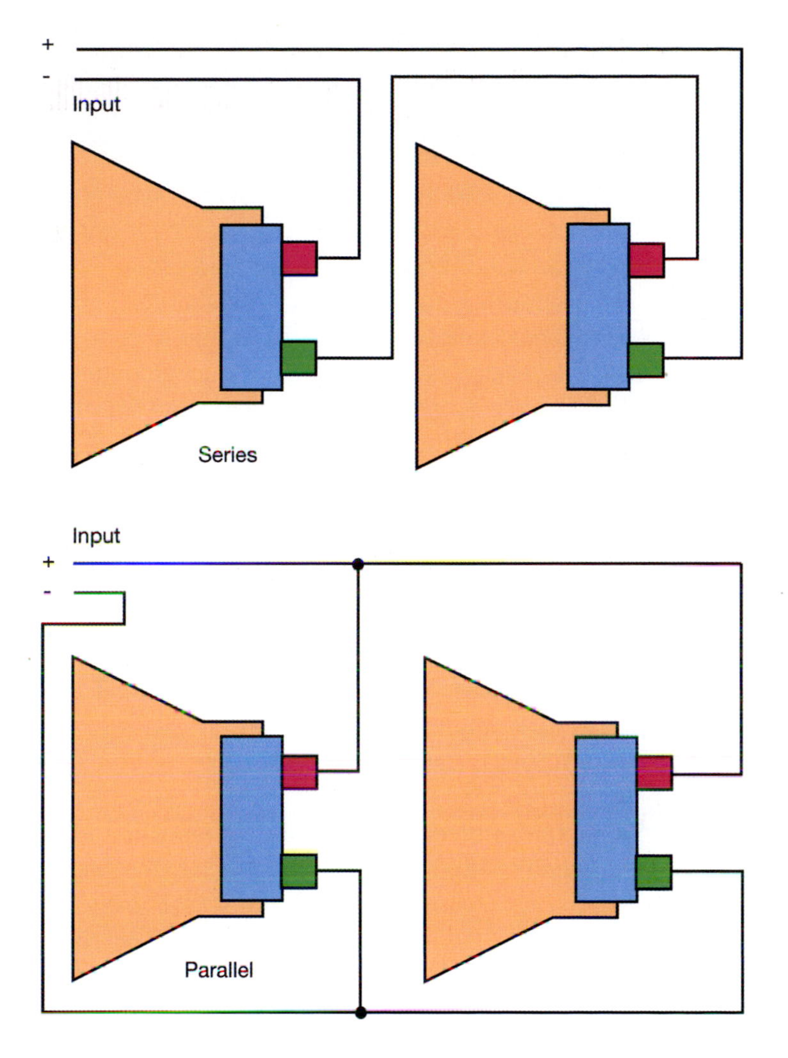

Because loudspeakers have a fairly low impedance, typically between four and 16 ohms each, it is essential that the cable used to connect them has an extremely low resistance, ideally a few hundredths of an ohm. If, for example a four-ohm speaker were to be connected to an amplifier by a long cable also having a resistance of four ohms, then half the power would be dissipated in the cable and half in the speaker—not a good use of amplifier power! That's why lightweight speaker cables or instrument leads should never be used, other than in dire emergencies. Low-impedance speaker cables tend to be much heavier gauge than signal cables and need not be screened.

Gain Structure

In broad terms, achieving the correct 'gain structure' means ensuring that every device in the signal chain is fed with a nominal signal level that fits into its 'comfort zone'. As you now know, if an electrical signal exceeds the level that the receiving device's circuitry can accommodate, distortion will result. Conversely, if the signal falls too far below the optimum operating level, the amount of amplification required to bring it back up to a usable level will also bring up any background circuit noise by the same amount, which will add more hiss than is necessary.

TIP: Proper gain structure means that every device in the signal chain should be fed with a nominal signal level that fits into its comfort zone.

While line-level signals are already within the ideal working range of most line inputs, extra care has to be taken when setting the gain of mic preamps on the console, as the output from a microphone can vary dramatically, depending on its sensitivity and the level of the sound it is picking up. By setting the input trim correctly, the signal level from the mic will be boosted to a level where the mixer's internal circuitry is able to deal with it without the risk of clipping distortion or added noise.

The mixer's PFL buttons provide an easy way to monitor and adjust individual channel-input levels, although you can also get close by setting the channel and output faders at their unity gain positions (the 0dB or U mark) and then feeding in the loudest signal you are likely to have to deal with. You can then tweak the gain trim so that the meters show a healthy level, with a safety margin before clipping.

Optimum Analogue Level

When optimising gain structure, we strive to keep the signal level fairly high while still leaving a useful safety margin to take care of unexpected peaks. Analogue VU meters have their 0VU mark where the red area starts; the safety margin or headroom is how far you can push the level beyond 0VU before distortion occurs and may typically be from 15dB to over 25dB. The actual figure should be shown on the specifications page of the manual.

With devices such as tape recorders and valve amplifiers, the amount of distortion rises gradually above 0VU, with hard clipping being the final stage once the headroom is all used up. The gentle distortion that occurs prior to hard clipping is often associated with analogue warmth, which is why recording engineers sometimes run their tape machines 'into the red' on purpose.

Setting Digital Levels

Digital circuits have similar limitations to analogue ones, in so far as a signal that is too high will clip, and one that is too low will cause background noise to become audible. However, there is no area of progressive distortion, so everything will sound clean right up until you hit the clip level, when it will sound very distorted indeed. Digital meters don't usually have an equivalent marking to the analogue 0VU, and so you have to decide on what nominal level to work at to leave sufficient headroom. 24-bit audio systems have exactly the same dynamic range as the best analogue systems, and so background noise isn't an issue even if you leave 20dB of headroom.

Gain And External Devices

We've covered using the mixer meters or PFL buttons to help you set individual input gain trim controls, but if you have any external effects or processors connected you also need to optimise levels through them. For example, if you have your console master effects send controls set too low, you'll need to add extra gain at the input of the FX device, and so you will end up with more noise than if you had the console send levels set higher to send a more healthy signal into the external device. Modern equipment is reasonably tolerant, though, and if you end up with both your output and input gain controls set at somewhere close to halfway you won't be far wrong. Where the external device doesn't have an adjustable gain trim, or it has a marked unity gain setting on its gain control, then simply turn up the send level on the mixer until the external device's meter shows a healthy reading while leaving adequate headroom.

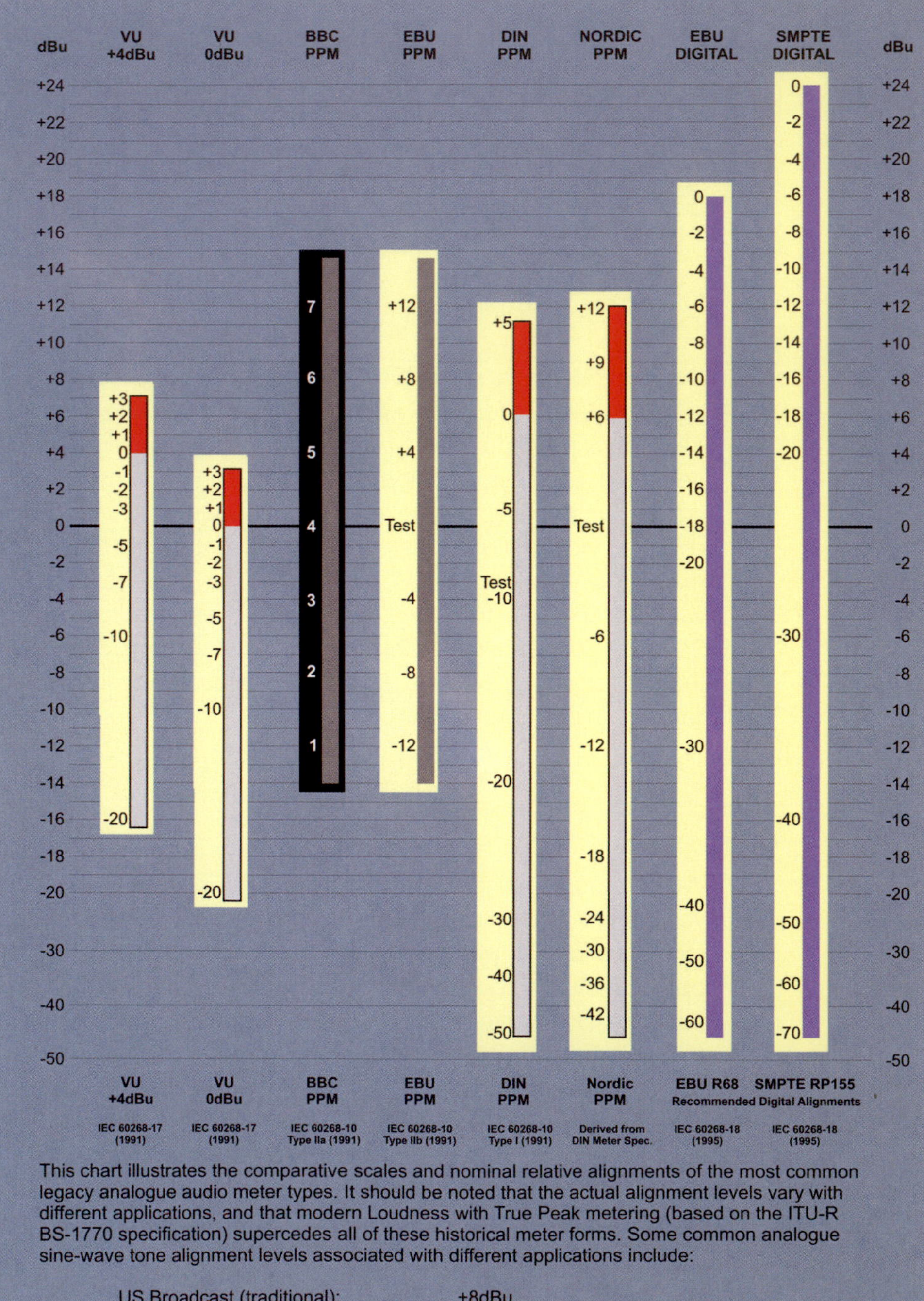

This chart illustrates the comparative scales and nominal relative alignments of the most common legacy analogue audio meter types. It should be noted that the actual alignment levels vary with different applications, and that modern Loudness with True Peak metering (based on the ITU-R BS-1770 specification) supercedes all of these historical meter forms. Some common analogue sine-wave tone alignment levels associated with different applications include:

US Broadcast (traditional):	+8dBu
US Music, Film and modern broadcast:	+4dBu
UK Broadcast:	0dBu
German Broadcast:	-3dBu
Other Territories:	Local Preferences

The fundamental difference between digital metering and analogue metering is the source of some confusion about optimum operating levels in digital systems.

On-stage Power

As discussed in Chapter 7, mains distribution boards and reels are a mainstay of the gigging band, and it is important that they are of decent quality since the contacts in the cheaper ones tend to work loose with repeated use. This can result in sparking, which can create loud, spluttery interference. In some cases you may be able to dismantle the offending power strip (making sure the other end is unplugged first!) and use pliers to pinch the contacts together to tighten them a little, but if the board is fully sealed you'll just have to discard it and buy another.

Always fully uncoil cable reels to prevent excessive heating, and make regular checks to ensure your power cables are in perfect condition. It also makes sense to buy heavy-duty cable reels as you won't be limited in what you can plug into them as long as you don't exceed the fuse capacity or the current-rating of the power point into which it is plugged. The maximum power you can draw from a UK 13-amp power point is a shade under three kilowatts, so simply add up the power ratings of the equipment you're plugging in to make sure you don't exceed this. Better still, restrict it to a couple of killowatts just to be on the safe side. In the USA, a 110-volt, 15-amp socket will safely supply a maximum of 1.8 kilowatts—in other countries you can

At the very least, you should perform a periodic basic visual safety inspection of any mains-power related cabling, looking for physical damage or anything that has worked loose.

calculate the maximum power you can draw from a socket by multiplying the mains voltage by the current rating of the socket. That will give you the power in watts. Power amps and lighting tend to take the most current, although modern LED lighting takes much less current than old-school incandescent bulbs.

A can of a specialised contact cleaner, like Caig's Deoxit D5, can sometimes be all that is needed to restore an audio circuit to life, especially in live sound gear that may have received only intermittent usage.

Even high quality units can suffer from oxidisation, which increases the contact resistance and can eventually lead to intermittent operation, so giving your mains connectors the occasional spray with a specialised contact treatment—such as Caig Deoxit D5—will help maintain peak performance. Deoxit also works wonders on audio connectors, and I've used it on more than one occasion to fix a noisy guitar jack prior to a gig.. It also makes an effective impromptu pot cleaner for noisy pots and sliders, even though it isn't designed for that purpose, so it's worth carrying a can in your toolbox. Traditional oily contact cleaners are to be avoided as they can cause dust and dirt to build up on the connectors.

When making or repairing your own mains cables, choose mains plugs with an effective cable clamp system—the cheaper ones that often use a brown fibreboard strip to secure the cable are prone to coming loose. IEC connectors are usually moulded onto their cables and so rarely cause problems. In the UK you should also fit the appropriate value mains fuse in the plug to suit the current rating of the cable. Most IEC cables are rated for 10A, but some are only good for 6A. You can use a colour code to identify cables with different fuse ratings by sticking bands of coloured insulation tape around the ends. It's also worth carrying some spare fuses for the equipment itself, paying attention to the size, type (fast or slo-blow), and current rating as these parameters will vary dramatically for different devices.

TIP: If in doubt about the total power consumption of your system, add up the power consumption figures (in VA or watts) for each device. The appropriate number should be listed on a plate near the mains connection on the units, and in their handbooks. Just make sure that the total comes to less than the maximum power rating for the mains wall socket. In the UK a standard 13-amp mains outlet can supply up to 2990 watts of power (13A x 230V), whilst a USA wall socket is good for between 1800 watts (15A x 120V) and 2400 watts, depending on whether it is a 15- or 20-amp outlet.

Generator Generalisations

When working outdoors with generator power, the grounding isn't always as effective as it should be so I'm in the habit of taking an earth spike with me that can be connected to the generator ground terminal. This needs to be hammered into damp ground, so if the weather has been very dry a few buckets of water will help ensure a low ground resistance.

In an ideal world, audio gear should be fed from its own generator—sharing with other equipment can cause problems, especially with the new generation of switch-mode amplifiers where the protection circuits are very fast to react. Anything that causes sudden current surges, such as beer pumps or tea urns, for example, can cause quite significant voltage drops lasting only a few milliseconds, but that's sometimes enough to trip the amplifier protection circuits, leaving you without audio for several seconds. Old-school analogue gear with heavy linear power supplies is less susceptible to such problems, as its reservoir capacitors are usually big enough to smooth out these short duration voltage dips.

Cable Types

Audio cables may be screened using various types of material such as woven copper braiding, spiral-wrapped multi-strand wire, metal foil, or conductive plastic. Each type of cable has advantages and disadvantages, and in addition to the effectiveness of the screening, you need to take into account handling noise, proclivity to kinking, and flexibility. For example, conductive plastic screened cables are resistant to kinking, have low handling noise and are easy to wire up, but are not as well-screened as cables with a woven screen. However, woven screens can exhibit high handling noise and may kink easily. Multi-strand double-wrapped screens offer the best of both worlds, but can be expensive.

Multicores

The multicore cable used to feed mic signals from the stage to the mixer comprises a number of small diameter, individually screened, twin core (balanced) cables, constrained within a common outer flexible sleeve. Multicores are used both in fixed installations and for connecting the mixing console to the stage box in conventional PA applications. Cable quality is an issue as cheaper multi-cores may fracture internally when repeatedly bent or trodden on, whereas better ones tend to be more durable. The way the multicores terminate at the stage-box end is also important as extreme bending at this junction must be

Multicores that fan into individual tails allowing the flexibility to repatch at the desk end, regardless of which lines the signals are plugged into on the stagebox.

avoided. Usually there's an additional flexible outer sleeve to stiffen the last 150mm or so of cable as well as an effective strain-relief gland.

At the other end, where the cable fans out into a bunch of individual leads terminating in male XLR plugs to connect with the mixer, the leads are identified with numbered sleeves corresponding to the stage box connectors, and there may also be a strain relief wire loop fixed to the end of the cable. This can be hooked onto a fixing point on the desk supporting the mixing console so the cables themselves don't take the weight of the multicore hanging off them.

Speaker Cables

Speaker cable must provide a low resistance path between the amplifier and the loudspeaker, to avoid power loss and a reduction in damping factor. These cables are not screened and usually have two cores, though some speaker systems require four or more cores where the woofers and HF drivers are fed from separate amplifiers. Damping factor is the output impedance of the amplifier divided into the impedance of the speaker connected to it, and the higher the damping factor, the greater the amplifier's ability to control the movement of the speaker cones. This in turn leads to a tighter sound, especially at the bass end of the spectrum.

The mantra here is to use the shortest, heavy-duty speaker leads that are practical, and use good quality, low resistance connectors such as Speakons, if your system has the option to use them. This will minimise power loss and maintain the best possible damping factor. Where jack leads must be used, try to source good quality cables where the connection to the tip of the jack plug is welded to the terminal connecting it, not riveted. Riveted jacks often work loose over time, which results in unnecessary extra resistance and, ultimately, intermittent operation. There's a limit to the thickness of cable you can fit into a jack plug but, again, use the best you can get. Also make sure jack-to-jack speaker leads are properly marked so they don't get mixed up with your screened instrument cables.

Glossary

0dB FS—Digital Full Scale, the level above which clipping will occur.

AC—Alternating Current. (See DC)

Acoustic Feedback—Continuous howling or whistling sound caused by the sound from a loudspeaker being picked up by a microphone and re-amplified with a loop gain of greater than unity.

Acoustics—A general term referring to the way sound behaves in a particular space such as a room or concert hall.

Active—A circuit that uses electrical power to amplify or otherwise process signals. Examples include power amplifiers, electronic crossovers designed to process line-level sources and effects units.

A/D or **Analogue-to-digital Converter**—Circuit for converting analogue waveforms into a sequence of equally-spaced samples, each of which has a numerical value corresponding to the instantaneous signal voltage. For the sampling process to work correctly, the incoming audio frequency must not exceed half the sampling frequency.

AES3—A two-channel, balanced interface used to transmit digital audio between devices, normally using 3-pin XLRs. The signal carries two digital audio channels plus clocking data at resolutions up to 24-bit.

AFL—After Fade Llisten, a means of monitoring a signal within a mixing console post-fader. In other words, the fader position also affects the level of the AFL monitored signal.

Algorithm—A software program designed to perform a specific task. For example, a digital reverb plug-in is based on a reverb algorithm. More complex software may comprise multiple separate algorithms, which can be considered as being software building blocks.

Aliasing—Refers to the process by which ambiguous data is generated when a waveform is sampled at less than twice the rate of the maximum frequency being sampled. Aliasing results in non-harmonically-related frequencies being added to the audible signal that can sound like whistles or metallic distortion.

Ambience—A term used to describe the effect of sound reflections in a space, such as a room. Where the decay time isn't long enough to produce an obvious reverb decay these reflections still affect the tonal character of the room.

Amp—Ampere, the standard unit of electrical current.

Amplifier—an electronic circuit capable of adding gain to an electrical signal.

Amplitude—Term describing the level of an electrical signal or sound.

Analogue—A system that handles electrical signals in the form of continually changing voltages or currents. (cf. Digital)

Anti-Aliasing Filter—A filter used to treat an analogue waveform prior to A/D conversion to ensure that no frequencies higher than half the sample rate are allowed though.

App—A program designed to run on a portable device such as smartphone or tablet.

Application—A term for a computer program.

Attack—Time taken for a sound to achieve maximum amplitude. For example, percussive sounds have a fast attack whereas bowed strings have a slower attack. In the context of compressors and gates, the attack time is the time taken for the device to respond once a signal reaches the threshold level.

Attenuate—To reduce the level of an electrical signal.

Audio—Airborne vibrations occurring within the range of human hearing, generally specified as being from 20Hz to 20kHz.

Aural Exciter—A device that adds upper harmonics to an audio signal to enhance and brighten it. The Aural Exciter name is a trademark of inventors Aphex.

Automation—A means of recording time-stamped fader moves, or other parameter changes, that can then be automatically played back in synchronism with the song to automate the desired changes. Automation tends to be found on digital mixers and is used in live performance to recall static snapshots or scenes comprising all the relevant mixer level and EQ settings.

Auxiliary or **Aux**—System used on a mixing console for sending pre- or post-fade channel signals to external monitoring systems (pre-fade) or to effects (post fade).

Aux Return—Dedicated mixer inputs used to feed the outputs of external effects units back into the mix.

Back-electret—Type of capacitor mic where a permanently charged material is fixed to the back-plate of the microphone capsule. (See Electret Microphone.)

Backup—Duplicate of digital data (often on hard drives, memory sticks or writable optical discs) for safety purposes.

BackLine—On stage instrument amplification such as guitar and bass amplifiers.

Balance—The relative levels of signals in the left and right channels of a stereo mixer channel or stereo mix. Mixers with stereo channels have a Balance control in place of the usual Pan control. Balance may also refer to the relative levels of the individual channel sources in a mix.

Balanced Connection—A system based around two conductors enclosed by a common screen, with a differential receiving device to reject interference. Both the sending and receiving device must have balanced I/O for the noise cancelling aspect of the connection to work.

Band—When used in the context of frequency, band often refers to a section of the audio spectrum such as that affected by a mid-range equaliser.

Band-pass Filter—Attenuates frequencies above and below the filter's pass-band frequency while passing or boosting those within the pass-band.

Bandwidth—The range of frequencies that a device can pass, measured between the -3dB points.

Bass Bin—A type of loudspeaker enclosure, often physically large and usually incorporating some form of horn-loading, designed to reproduce bass frequencies at high SPLs.

Bass End—A very general term for the low end of the audio spectrum from 20Hz to around 120Hz.

Bell Equaliser—Cuts or boosts frequencies within a particular frequency range. The frequency response diagram is roughly bell-shaped, hence the name.

Bi-amped—When related to loudspeakers this term means using separate amplifiers to power the HF (tweeter) and LF (woofer) drivers. The two amplifiers would normally be fed from an active crossover circuit.

Binary—Mathematical system used in computers where the data comprises ones and zeroes. This is well suited to switching circuitry which has only two states, on or off.

Birch Ply—A high quality type of plywood often used in loudspeaker cabinet construction.

Bit—Individual element of binary digital data comprising a one or a zero.

Boundary Effect—At very low frequencies, a bass speaker is almost omnidirectional so by placing it very close to the wall, all the reflected sound from the rear of the cabinet will be reflected virtually in-phase with the sound from the front of the cabinet, resulting in a near doubling of level.

BPM—Beats Per Minute.

Buffer—Circuit designed to isolate a device's output from the rest of the circuitry and to present a low impedance to the destination device.

Bus—In the context of mixing, a bus is a signal path along which signals travel and into which other signals may be mixed. A mixer includes buses for the stereo mix, any mix sub-groups, the various aux sends, and for the channel solo/PFL signals.

Byte—A collection of digital data made up of eight bits.

Cable Tester—Device designed to test cables terminating in popular audio connector types such as jack, XLR, phono and Speakon. The usual approach is to use a series of LEDs to warn of incorrectly connected cables, open circuits due to faulty connections, or broken conductors, and short circuits.

Capacitor—Electronic component capable of storing an electrostatic charge. The impedance of a capacitor reduces with increasing frequency.

Capacitor Microphones—Microphones that work on the principle of changing electrical capacitance between a stationary electrode and a moving, electrically conductive diaphragm. Capacitor microphones require power to polarise the capacitive pickup element and to run the internal impedance-converting preamplifier.

Capsule—That part of a microphone that converts sound into an electrical signal.

Cardioid, also **Unidirectional**—A term usually applied to microphones designed to pick up sound over a relatively narrow angle. The polar pattern is approximately heart-shaped, hence the term cardioid.

Channel (Mixer)—That part of mixer that handles an input signal prior to it being fed to a mix bus. These typically include controls for adjusting gain, EQ, send levels, routing and overall output level.

Chip—Integrated circuit—an active electronic device.

Chorus—Effect created by adding a pitch modulated, delayed version of a signal to the original.

Chromatic—A music scale comprising semitone intervals.

Click Track—Regular audible click used to help musicians to keep in time with pre-recorded backing parts.

Clipping—Distortion that occurs when a signal attempts to exceed the maximum level a piece of equipment can accommodate. As the signal amplitude can never exceed this level, the tops of the waveform become flattened or clipped.

Coaxial Screened Cable—Cable comprising a central wire surrounded by a conductive, tubular screen, usually fabricated from woven metal, metal foil or conductive plastic. This outer conductive sheath provides a path by which interference can be drained away to ground. An insulating outer sheath is normally fitted.

Comb Filter—The filtering effect created when an audio signal is added to a slightly delayed version of itself. Some frequencies add while others cancel, producing a response with many peaks and dips, rather like the teeth of a comb.

Common Mode Rejection—A means of specifying how effectively a balanced circuit rejects a signal that is common to both the hot and cold balanced input conductors (usually noise and interference).

Compander—A system of compressing an audio signal to reduce its dynamic range before sending it through a transmission system such as a radio link. An expander in the receiver, having the inverse characteristics of the compressor, restores the original dynamic range while reducing the level of background noise introduced by the transmission system.

Compression Driver—Specialised high-frequency loudspeaker driver used in conjunction with a horn flare where the narrowest point of the horn (throat) is smaller in diameter than the diaphragm of the driver.

Compressor—Signal processor designed to reduce the dynamic range of audio signals. The usual mode of operation is that signals exceeding a threshold set by the user are reduced in level by an amount determined by the ratio setting.

Condenser—Old-school term for capacitor.

Conductive Plastic—A specially formulated plastic designed to be electrically conductive. It is used in some high quality faders and also as a means of screening cables where flexibility and resistance to kinking are important.

Conductor—Material that facilitates the transmission of electrical current.

Console—Alternative term for mixing desk.

Constant Directivity—Specially shaped high-frequency horn intended to maintain a constant dispersion pattern across its frequency range.

CPU or **Central Processing Unit**—That part of the computer than manipulates data.

Critical Distance—The distance from a sound source in an enclosed space at which the direct and reverberant sound levels are equal.

Crossover—An electrical or electronic circuit designed to split the audio spectrum into two or more regions, for example to route only high frequencies to the tweeters and low frequencies to the bass speakers or woofers. Crossovers may be passive, in which case they split the output from an amplifier carrying a full-range audio signal to feed the separate drivers, or they may be active, in which case they split the signal before feeding it to individual amplifiers directly powering the various drivers.

Cut-off Frequency—Often associated with high- and low-cut filters, the cut-off frequency is the point above or below which attenuation begins, often measured at the –3dB point.

Cycle—One complete vibration of a sound wave or electrical signal. In the 1970s the very logical term 'Cycles per Second' was replaced by Hertz.

Damping—As applied to acoustics, damping describes the way sound energy is absorbed to control reverberation time. Damping is usually different at different frequencies in a real life environment where high frequencies tend to decay more quickly than lower frequencies. In an artificial reverberation device damping describes an adjustable parameter that emulates the way damping occurs in a real environment.

Data Reduction—A means of reducing the amount of data needed to represent a digital audio signal. Consumer audio devices use systems such as MP3, although data reduction is also used in digital radio mic systems.

DC—Direct Current—the unidirectional flow of electric charge, i.e. not alternating in polarity. (See AC)

Decay—A reduction in amplitude over time such as occurs when a plucked string loses energy.

Decibel or **dB**—Unit used to describe the ratio of two electrical voltages, powers or audio levels.

dBm—describes the signal level obtained when a power of 1 milliwatt is dissipated into a 600 ohms load. 0dBm corresponds to a signal voltage of 0.775 volts, but is no longer used in the context of audio systems as it applies exclusively in 600-ohm impedance-matched systems.

dBu—(also dBv) A standard signal level format for modern analogue audio systems. 0dBu represents a signal voltage of 0.775 volt, but without the requirement for a 600 ohm load. Used with constant-voltage audio interface formats. +4dBu is a common standard operating level for professional analogue equipment.

dBV—Another standard signal level format for constant-voltage audio interfaces. 0dBV represents a signal voltage of 1 volt. 10dBV is a common standard operating level for semi-pro and consumer equipment.

De-Esser—Device for reducing the level of sibilance (S,T and F sounds) in vocal signals. A de-esser is basically a frequency selective compressor that attenuates the signal whenever high levels of sibilant frequencies are detected.

Detent—Physical click position to denote the centre of a rotary control, usually where the centre position is the default for no action, as it would be in the case of a pan control or EQ cut/ boost knob.

DI Box—Device which accepts the signal from a guitar, bass, or keyboard and conditions it to resemble a microphone signal at its output. The output is a mic-level, balanced signal with a low source impedance, capable of driving long mic cables. Usually includes a ground-lift facility to avoid ground loop noises. Both active and passive versions are available, the former requiring power from internal batteries or phantom power via the mic cable. Active DI boxes generally have higher input impedances than passive types and are generally considered to sound better.

Digital—A system that handles electrical signals in the form of discrete-time measurements of a signal's amplitude, coded in a binary form. (cf. Analogue)

Digital Delay Line (DDL)—Used for various purposes, including the creation of echo and modulation effects. In large scale sound systems DDLs are also used to compensate for the acoustic time delay that occurs between speakers set up at different distances from the listener.

Digital Device—In the context of audio, a digital device is one that processes a signal that has been converted from an analogue format to a string of binary numbers representing instantaneous waveform measurement values made at twice the audio frequency or above.

Digital Full Scale—Denoted by dB FS. This is the maximum level a digital system can handle before the signal is clipped.

Direct Injection or **DI**—A technique for recording an electrical signal directly, for example, the output from a bass amp, the pickup of a guitar, or from an electronic keyboard.

Digital Reverberator—Digital device for producing a reverberation effect simulating the behaviour of sound in a real space such as a room or hall. The two common methods are algorithmic (synthetic), where delays and feedback loops are used to simulate the effect of natural reverb, and convolution, where an impulse response recorded in a real space is used to impose the acoustics of that space onto an electrical signal.

Digital Speaker Controller—A combination of crossovers, equalisers, delays and limiters designed to achieve optimum performance from a PA loudspeaker system while protecting it from overload.

Directivity—Describes the angle of coverage of a loudspeaker system, both in the vertical and horizontal planes. Higher directivity equates to a narrower angle of coverage.

Dispersion—A means of describing the angular width and height of a loudspeaker system's coverage.

Dither—The process of adding low level noise to a digitised audio signal in such a way as to extend its low level resolution and reduce quantisation distortion. Though dither increases the level of noise very slightly, it allows very low level audio signals to remain undistorted.

Driver—Term describing a loudspeaker chassis before it is fitted into a cabinet.

Dry—Signal that has had no effects added.

Dynamic Microphones—Microphones that work on the electromagnetic principle with a coil of wire, attached to a thin diaphragm, moving in a magnetic field.

Dynamic Range—The range between the highest signal that can be handled by a piece of equipment with clipping at the upper extreme and the level at which small signals disappear into the noise floor at the lower extreme.

Dynamics—Term describing how the levels change throughout a piece of music.

Early Reflections—The initial sound reflections from walls, floors, ceilings or other solid objects following a sound created in an acoustically reflective environment. These re-reflect and quickly build up into a dense reverb decay.

Effect—Usually describes a process that involves adding the dry signal to a modified version of itself, (usually involving an element of delay) such as reverb and echo. Can also describe any process that changes the character of a sound in a creative rather than corrective way.

Effect Return—A dedicated mixer input designed to accept the output from an effects unit (or its virtual DAW equivalent).

Effects Loop—A system of using an aux send to feed an external effects unit or plug-in which is then routed back into the mix via an aux return or spare input channel.

Electret Microphone—Type of capacitor microphone that uses a permanently charged capsule.

Encode/Decode—A system that requires a signal to be processed prior to recording or transmission, then the reverse of that process applied during playback. Companders used in radio mic systems are examples of encode/decode processes.

Enhancer—A device capable of brightening audio material using a variety of techniques that go beyond conventional EQ. These may include dynamic equalisation, phase shifting and harmonic generation. The Aphex Aural Exciter is probably the best known example.

Envelope—Describes the way the overall level of an electrical signal changes with time.

Envelope Generator—Circuit or software designed to generate a control signal which in turn controls the level of an audio signal to impose a level envelope upon it. The side-chain circuit of a compressor or gate generates a 'control' envelope based on the input signal and on the attack and release settings.

Equaliser—Device for selectively cutting or boosting the levels of selected parts of the audio spectrum. These vary in complexity from simple bass and treble controls to multi-band parametric designs.

Exciter—See Enhancer.

Expander—A dynamic control circuit or plug-in designed to decrease the level of low level signals, rather like a compressor in reverse. The result is an increase in dynamic range of the signal being processed.

Fader—A potentiometer controlled by a slider.

Faraday's Law—Any change in the magnetic field surrounding a coil of wire will cause a voltage (emf) to be induced in the coil. Conversely, when an electrical current is passed through a coil placed in a magnetic field, a force (thrust) will be generated between the coil and magnetic field.

FET—Field Effect Transistor: type of solid-state semiconductor.

Figure-of-eight—The polar response of a microphone where the diaphragm is open to the air on both sides. These are equally sensitive both front and rear but reject sounds coming from 90 degrees off-axis.

File—A collection of digital data stored so as to appear as a single item.

Filter—Electronic circuit that can cut a specific range of frequencies, or frequencies that occur above or below a specified cut-off point.

Flanging—Originally created by running two tape recorders slightly out of sync, the effect is now more commonly created using a modulated delay line with feedback. The delay time is usually only a few tens of milliseconds so the resulting comb-filtering creates a distinctive sweeping sound.

Flat Frequency Response—Describes the performance of a system which applies an equal amount of gain to all frequencies. In practice, a small amount of deviation is permitted within specified limits.

Flash Memory—A type of solid state memory that retains data even when the power is removed. Flash memory is used in tablet-style computers, smart phones, many hand-held recorders and so on, but is also making inroads into mainstream computing as its capacity increases and costs fall.

Fleming's Left Hand Rule—A way to remember the relationship between thrust, magnetic field direction and current, where the upward pointing thumb of the left hand represents 'thrust', the forward-pointing forefinger 'field' and the second finger pointing to the right 'current'.

Flutter Echo—A resonant echo that occurs when sound reflects back and forth between two parallel, reflective surfaces. The closer the surfaces, the higher pitched the resonance.

Foldback System—See Stage Monitoring system.

Folded horn—A means of reducing the size of a horn-loaded bass speaker by using baffles to create a labyrinth that gradually increases in cross-sectional area to form a horn that is, quite literally, folded.

Formant—A fixed frequency component or resonance that defines the character of instrument or voice. For example, the body resonance of a violin or acoustic guitar.

Format—Can describe a file type, such as WAV or MP3, but is also used to describe the process of initialising a digital storage device such as a hard drive to make it compatible with the connected computer. Different computers sometimes use different formatting systems.

Frequency—Measure of how many cycles of a repetitive waveform occur in one second. (See Cycles.)

Fundamental Frequency—Usually the lowest frequency component of a sound, such as the pitch of a vibrating string. Most instruments produce a fundamental frequency plus a series of harmonics and partials at higher frequencies.

FX—Short for Effects.

Gain—The amount by which a device amplifies a signal, usually expressed in dB.

Galvanic Isolation—A system for coupling electrical signals without requiring direct electrical contact, examples being audio transformers (which couple via interacting magnetic fields) and opto-isolators that convert electrical signals to modulated light signals and vice versa.

Gate—A device designed to attenuate or mute low level signals falling below a user-defined threshold so as to attenuate noise during pauses.

General MIDI or **GM**—part of the current MIDI specification that defines a sound set to assure a minimum level of compatibility when playing back GM MIDI song files. The GM spec defines which sound types correspond to which program numbers, minimum levels of polyphony and multi-timbrality, response to controller information and other factors.

Glitch—Often used to describe a momentary corruption of a signal (such as a click or gap), or unexplained software/hardware behaviour.

Graphic EQ—An equaliser able to adjust several narrow and adjacent regions of the audio spectrum using individual cut/boost faders. The fader positions provide a graphic representation of the EQ curve, hence the name.

Ground—Electrical earth; the ground cable in a mains power system is physically connected to the ground (or earth) via a conductive metal spike that forms part of the building's wiring system. The metal cases of equipment are usually connected to ground via the safety ground conductor of the mains cable.

Ground Loop—Where multiple ground connections exist between pieces of audio equipment connected together, audible mains hum may result due to circulating current induced into the resulting loops. Ground loops are sometimes also referred to as earth loops.

Group—Describes the method of sub-mixing specific mixer channels by routing them to separate mix buses in a mixing console to allow multiple related sources, such as the mics over a drum kit, to be controlled using a single fader. Group may also be used to describe a number of mixer channel faders 'grouped' together so that adjusting any one fader in the group adjusts all of them—a function often available in large format digital mixing consoles.

Harmonic—High frequency component of a complex waveform that is a multiple of the fundamental.

Harmonic Distortion—Harmonics added when a waveform shape is changed by a distorting mechanism, such as clipping or a non-linear circuit.

Headroom—A safety margin (usually expressed in dB) between the nominal operating range of a signal (often expressed in analogue systems as 0VU) and the maximum level the equipment can handle before clipping. Digital systems have no inbuilt headroom as clipping occurs as soon as the signal reaches 0dB FS so the operator has to leave an appropriate amount of headroom by setting a suitable nominal level well below 0dB FS.

Hertz—(Abbreviated to Hz) The unit of frequency that describes the number of cycles per second of a repeating waveform in one second.

High-pass Filter (HPF)—A filter that attenuates all frequencies below its cut-off frequency.

Hiss—Noise caused by random electrical fluctuations in electronic components.

Horn-loaded—A means of increasing the efficiency of a drive unit by using a horn to more closely match the acoustic impedance of the driver to the air mass that it couples with.

Horn tweeter—A high-frequency loudspeaker driver which has a horn-shaped flare fixed to the front in order to increase the acoustic efficiency and better control the directivity.

Hum—The unwanted addition of low frequencies related to the mains power frequency.

Hyper-cardioid—refers to a form of unidirectional microphone with a narrower pickup angle than a standard cardioid pattern, and a small rearward pickup lobe.

Impedance—A circuit's resistance to alternating current, which varies with frequency.

Impedance Matching —An interface technique in which the output of one device passes the maximum amount of power to the input of the next because the source and load impedances are identical. Employed mainly in high-frequency interfaces such as video and digital audio. It can also relate to the mechanical matching of a drive unit's cone motion to the air it is required to move, often by means of a shaped flare or horn.

Inductor—A reactive component, usually a coil, where the impedance increases with increasing frequency.

In-ear Monitoring—(IEM) The use of in-ear phones in place of on-stage monitor loudspeakers, often fed via a radio system.

Insert Point—A connection point in a mixer or other device that allows an external processor to be patched or 'inserted' into a signal path so that the signal flows through the external processor.

Interface—Generically, a means of passing signals between devices. Usually refers to audio or MIDI interfaces that allow computers to send and receive audio or MIDI signals. Some interfaces include both Audio and MIDI capability and may connect to the computer via Firewire, USB, Thunderbolt or via a network connector.

Intermodulation Distortion—A form of distortion that introduces frequencies not present in the original signal based on the sum and difference products of the original frequencies.

Inverse Square Law—A point source of sound generates a spherical wavefront where the sound intensity diminishes in proportion to the square of the distance from the speaker.

Jack—Commonly used audio connector. May be mono or stereo and available in consumer 'mini' formats and the more common quarter inch type used in music applications. There are also specialised Bantam and Post Office jacks used in some professional patchbays but these are not compatible with other types of jack.

Jitter—A measure of the instability in a digital clocking system, usually measured in parts per million.

Kilo or **k**—Abbreviation for 1000. 1kHz, therefore means 1000Hz.

LCD—Liquid Crystal Display, often used as displays for synths and FX processors.

LED—Light Emitting Diode, a common type of solid state lamp that emits very little heat.

LFO—Low Frequency Oscillator, often used as a source of modulation to create effects such as chorus or vibrato.

Limiter—A device that limits the maximum amplitude of an audio signal to prevent it from exceeding a pre-defined level.

Line Array—A type of system that uses vertically stacked drivers or cabinets to produce a wide horizontal dispersion and a narrow vertical dispersion. This is due to the production of near-cylindrical wavefronts rather than the spherical wavefront of a 'point'source' speaker.

Line Level—This is a slightly vague term and applies to non-microphone signals with a peak level measured in volts rather than millivolts. Nominal line level is around -10dBV for semi-pro equipment and +4dBu for professional equipment.

Linear—A device or circuit that changes only the level of the signal being passed.

Load—An electrical or electronic circuit that draws power from the circuit feeding it.

Loop—A multipurpose term. Loop can describe a signal path where the output of a device is connected back to the input. (See Ground Loop). A loop can also mean a segment of audio that can be repeated continually in a musically meaningful way via a sampler, such as a bar of drum rhythms.

Low-pass Filter (LPF)—A filter that attenuates all frequencies above its cut-off frequency.

M and **m**—Lower case m denotes milli meaning one thousandth. Upper case M denotes Mega which means one million.

Magnetic Pole-piece—A structure made from a material of high magnetic permeability and usually attached to a magnet, used to direct its magnetic field.

Memory—Computer memory usually describes RAM memory used to store programs and data while the computer is running, though a hard drive or a 'flash' memory stick is, technically speaking, also a form of memory. RAM data is lost when the computer is switched off. Flash memory, however, retains its data when the power is switched off.

Mic Level—Describes the level available from a typical microphone, which must be amplified many times, perhaps by up to 60dB or more, to increase it to line level.

MIDI—Musical Instrument Digital Interface. MIDI is used to allow electronic instruments to communicate but may also form part of a controls system, for example, where a pedal unit capable of generating MIDI commands is used to control an effects device that has MIDI control capability.

MIDI Controller—Term used to describe the physical music instrument interface used to input a musical performance as MIDI data. Keyboards are the most obvious examples but there are also drum pads, wind synths, and MIDI guitars that output MIDI data.

MIDI In—The connection used to receive information from a master controller or from the MIDI Thru socket of another slave unit.

MIDI Note Off—MIDI message transmitted when note is ended or released on the MIDI controller.

MIDI Note On—MIDI message transmitted when note is played on the MIDI controller.

MIDI Out—Connector used to send data from a master device to the MIDI In of a connected MIDI device.

MIDI Port—MIDI is limited to 16 MIDI channels, but by using a MIDI interface with several independent MIDI ports, it is possible to have several groups of 16 MIDI channels operating at the same time. A multi-port MIDI interface must be supported by the host DAW software.

MIDI Sound Module—A hardware, MIDI-controlled sound generating instrument with no integral keyboard.

MIDI Thru—The connection on a slave unit used to feed the MIDI In socket of the next unit in line. This passes the MIDI In signal through without change.

Mix Engineer—Sound engineer experienced in mixing live music.

Mixer—A device that accepts a number of microphone and/or line-level signal sources and mixes them together to form a composite signal.

Mixer/Amplifier—A single unit containing both mixer and power amplifier(s).

Monitor—Speaker or in-ear system used to allow the performers on stage to hear the desired balance of mix elements to help with their own performance.

Monophonic—One musical note at a time with no overlaps or chords. Alternately, a single audio channel. (cf. Stereo)

Multicore—Sometimes referred to as a snake. A type of multicore cable used to connect a stage box to a remote mixer and comprising several individually screened cable pairs within a single flexible outer sheath.

Multi-timbral—The ability of a MIDI Sound Source (hardware or software) to produce several different sounds at the same time where each can be controlled by a different MIDI channel.

Nyquist Theorem—Named after the engineer who proved the rule that states that a digital sampling system must have a sample rate at least twice as high as that of the highest frequency being sampled to avoid aliasing.

Octave—A musical frequency or pitch that is double (or half) the frequency of the original.

Ohm—Unit of electrical resistance.

Omnidirectional—A polar pattern that is equally sensitive to sound regardless of the angle it approaches the microphone.

Op-amp—Short for Operational Amplifier. A type of amplifier often built using integrated circuit technology and having differential inputs. These are found in many audio devices including mixing consoles.

Open Circuit—A break in an electrical circuit across which current is unable to flow.

Operating System—The host software that enables a computer to load and run programs, common examples being Windows, Mac OS and Linux.

Opto Electronic—A device where an electrical parameter, such as resistance, is affected by a variation in light intensity, such as light falling on a photocell.

Oscillator—A circuit designed to generate a repeating electrical waveform.

Overload—Commonly used to describe the action of either feeding excessive signal levels into a device or applying so much gain that distortion (or in the case of a loudspeaker, also the risk of damage) results.

Pad—A resistive attenuator for reducing signal level, often found in microphones, DI boxes, mixer input stages and mic preamps.

Pan-Pot—Short for panoramic potentiometer: a control designed to steer a mono signal anywhere between the left and right mix buses to create the illusion of stereo positioning.

Parallel—In terms of signal flow, parallel means connecting two or more circuits together so that their inputs are fed from a common source, and their outputs are combined.

Parameter—A variable value, such as an EQ frequency, that affects the performance of an audio device or software.

Parametric EQ—An equaliser with separate controls for frequency, bandwidth and gain and which can usually apply both cut and boost.

Passive—A circuit with no active elements such as valves, transistors or ICs.

Passive two-way—Refers to a loudspeaker system comprising a bass driver and a tweeter fed from a passive crossover filter network comprising capacitors, resistors and inductors. Passive three-way systems are less common but do exist.

Patch—Term used to describe a means of storing and recalling the settings of an electronic instrument, effects unit or digital mixer. Mixer patches are often referred to as Scenes.

PFL or **Pre Fade Listen**—A means to monitor an audio signal prior to the fader controlling its level.

Phase—The offset between two sine waves of identical frequency expressed in degrees where 360 degrees corresponds to a delay of exactly one cycle.

Phaser—A common effect which combines a signal with a phase-shifted version of itself to produce tonal filtering effects. The phase shift is modulated by an LFO to create a cyclic tonal change.

Phono Plug—Sometimes called an RCA phono jack. A simple push-in, unbalanced Hi-Fi connector developed by RCA and used extensively on semi-pro equipment and also for connecting S/PDIF digital audio signals. Many mixers have a pair of phono connectors to allow the connection of a consumer audio playback device.

Pink Noise—A random signal with a power spectral density which is inversely proportional to the frequency. Each octave carries an equal amount of noise power. Pink noise sounds natural, and resembles the sound of a waterfall. (cf. White Noise)

Pitch—The musical interpretation of audio frequency where, in Western music, each octave is divided logarithmically into 12 semitones.

Pitch-Shifter—Device for changing the musical pitch of an audio signal without changing its duration.

Point Source—A speaker where all the sound can be considered as coming from the same point in space.

Polar Pattern—The sensitivity of a microphone as plotted against angle over a range of frequencies relative to the frontal axis. Loudspeaker cabinets also have a polar pattern that shows their directivity across the frequency range.

Polarity Invert—The correct term for a circuit that inverts an electrical signal, though the term phase (or the Greek symbol Ø) is often used in a somewhat sloppy way to describe the polarity invert switch found on mixer inputs and mic preamps.

Ported Enclosure—A speaker cabinet featuring a port or vent that allows the resonance of the cabinet to be tuned to improve the low frequency performance of the system.

Post-Fade Aux—An aux signal sourced after the channel fader so that the aux send level follows any channel fader changes. The most common use of a post-fade aux is to feed effects devices to maintain the same balance of dry signal and effect regardless of the channel fader position.

Power Amplifier—Amplifier designed to accept line-level signals and boost them to the high power levels necessary to drive loudspeakers. Most, (but not all) professional power amplifiers have two channels for stereo operation.

Power Compression—The drop in efficiency of a loudspeaker that occurs when the voice coil and magnetic system heats up.

Power Supply—A unit able to convert mains electricity to the voltages necessary to power a device such as mixer, valve microphone or amplifier.

PPM—Peak Programme Meter; a meter designed to register an approximation of signal peak levels.

Pre-Fade Aux— Aux signal taken from before the channel fader, usually for setting up foldback/monitor/cue mixes.

Preset—Effects unit or instrument patch provided by the manufacturer.

Pressure Gradient Microphone—A microphone that senses the difference in pressure between the front and rear of the diaphragm and which may have a figure-of-eight, cardioid, or hyper-cardioid polar pattern depending on the capsule design.

Pressure Microphone—A microphone that senses only local pressure changes and has an omnidirectional polar response.

Processor—Term used to describe an audio device designed to treat an audio signal by changing its dynamics or frequency content, and in most cases (other than the special cases of parallel compression and distortion) the whole signal passes through the device rather than being added to the dry signal. Examples of processors include compressors, limiters, gates, distortion devices and equalisers. While effects such as reverb and delay are created by signal processing, it helps avoid confusion if such devices are referred to as Effects. As a very general rule, processors are normally used via track, bus or master insert points whereas effects can be used both in insert points and in the aux send/return loop.

Proximity Effect—An increase in sensitivity to low frequency sounds from very close sound sources exhibited to some degree by all pressure gradient microphones.

Pulse Width Modulation—An audio signal can be represented by a series of equal amplitude pulses where the pulse width is varied according to the amplitude of the audio signal. Integrating this signal (a form of smoothing) reconstructs the original audio waveform. This technique allows very efficient high power amplifiers to be built with low weight and less heating and is employed in most Class-D amplifiers.

Q—An alternative to the term Bandwidth for measuring the resonant properties of a parametric equaliser. The higher the Q, the narrower the range of frequencies that pass through. In other words, a narrow bandwidth equates to a high Q value and vice versa where Q is defined as the filter's centre frequency divided by the bandwidth of the filter measured at the -3dB points either side of the centre frequency.

Radio Mic—Describes a wireless link between a microphone and receiver that uses radio waves. Radio guitar systems are also available that operate on the same principle.

Radio Spectrum —Essentially the entire range of radio frequencies used for media and data transmission where only a very small part is available to the wireless microphones operator.

RAM—Abbreviation for Random Access Memory. Data stored in RAM is lost when the power is turned off.

Release—The time taken for a level or gain to return to normal, for example, when the signal level falls below the threshold in a compressor. Release is also used to describe the rate at which a synthesised sound falls in level after the key has been released.

Resistance—A material's opposition to the flow of electrical current, measured in ohms. A good conductor has a low resistance whereas a poor conductor will have a high resistance.

Resonance—The ability of a tuned circuit or mechanical resonator to store and release energy at a specific 'tuned' frequency.

Reverberant—Describes the acoustic property of a space or room where the natural level of reflected sound gives rise to audible reverberation.

RF—Radio Frequency (nominally electromagnetic signals transmitted above about 50kHz).

Ribbon Microphone—A microphone employing a thin conductive ribbon suspended in a magnetic field. As the ribbon vibrates, a small electrical current is generated within the ribbon.

RMS—(Root Mean Square) A method of specifying the behaviour of a piece of electrical equipment under continuous sine wave testing conditions.

Roll-Off—The rate at which a filter attenuates a signal once it has passed the filter cut-off point and usually expressed in dB/octave.

ROM—Abbreviation for Read Only Memory. Rom contains data that can't be changed and the memory remains intact when the power is removed. An E-PROM—(Erasable Programmable Read Only Memory) operates in a similar manner to ROM, but the information on the chip can be erased and replaced using specialised equipment.

Rotary Speaker—In fact most so-called rotary speaker devices use fixed speakers firing into rotating baffles or horns to produce both pitch and timbral modulation.

Safe—Other than the obvious, Safe is often used to describe a digital mixer parameter that can be isolated or made safe so that it isn't changed when a new scene is loaded.

Sample Rate—The rate at which an A/D converter samples the incoming waveform.

Sawtooth Wave—Resembling the teeth of a saw, this waveform contains both odd and even harmonics.

Scene—Digital mixer patch used to save and recall a set of fader, send level, EQ and FX settings for a particular purpose.

Sequencer—A device for recording, editing and replaying MIDI data, usually in a multitrack format. Sequencers were the forerunners to today's DAWs.

Short Circuit—A low resistance path, usually undesirable, that allows electrical current to flow when a fault occurs.

Shorting Rings—Non-ferrous conductive rings placed in alignment with a loudspeaker's voice coil that act as a 'shorted' turn, with the aim of providing greater linearity over the travel of the voice coil.

Sibilance—High frequency whistling sound that affects some vocal recordings (caused by airflow around the singer's teeth), specifically on F, S and T sounds.

Side-Chain—That part of the circuit that follows the envelope of the main signal to derive control signals for processing, such as gain control in a compressor.

Signal—Electrical representation of a waveform, such as the output from a microphone, which follows the acoustic envelope.

Signal Chain—The entire route taken by a signal from the input to a system to the output.

Signal-to-Noise Ratio—The ratio usually expressed in decibels of maximum signal level (before clipping) to the residual noise.

Sine Wave—A pure tone with no harmonics.

Speaker Columns—Type of loudspeaker enclosure where the loudspeakers are arranged in a vertical row.

S/PDIF—Pronounced 'Spidiff' and standing for Sony/Philips Digital Interface, the S/PDIF digital data format is very similar to the professional AES3 standard although it is unbalanced and operates at a slightly different voltage level. It is a two-channel system and can accommodate up to 24-bits of audio data as well as track start flags, source identification information, and timing data needed by consumer CD players and other devices. Though the connectors are conventional RCA phonos (there's also an optical equivalent called TOSLink), the cable must be 75-ohm coax to operate reliably.

Spill—Unwanted leakage of sound from loudspeakers or other sources into live microphones.

SPL—Sound Pressure Level, expressed in decibels.

Square wave—A symmetrical rectangular waveform comprising a series of only odd harmonics.

Stage Box—A box fitted with multiple connectors, usually XLRs, enabling microphones to be fed into a multicore. Usually also includes opposite polarity connectors to allow mixer outputs to be fed back to the stage to drive monitors, amplifier racks or active speakers.

Stage Monitoring—A system of amplifiers and loudspeakers that enable the performers on stage to hear what they are playing and singing.

Sub Bass—very low frequencies below the range of typical main loudspeakers, typically 40Hz and below.

Sustain—a sound characteristic that describes how long the sound takes to decay after the initial note has been played.

Switched-mode power supply—A type of power supply that uses high frequencies and very small transformers to achieve the same performance as conventional transformers operating at 50/60Hz, but with greater efficiency and less weight.

Tempo—The speed or 'beat' of a piece of music measured in beats per minute.

THD—Total Harmonic Distortion.

Thru—An output socket on an audio device, such as a DI box, that passes through the input signal with no processing. May also refer to a MIDI connector which passes through the signal present at the MIDI In socket.

Timbre—The tonal 'colour' of a sound determined by its harmonic complexity and the way those harmonics vary with time.

Transducer—Any device for converting one form of energy to another, such as a microphone or loudspeaker.

Transient—Part of a sound, usually the start, where the harmonic content and level changes abruptly, such as the beginning of a drum hit or the onset of a picked guitar string.

Transparency—A subjective term used to describe clear, uncoloured audio quality that sounds true to the original source.

Transpose—To change the key of a musical signal or performance by a fixed number of semitones.

Tremolo—Cyclic modulation of amplitude.

TRS Jack—Stereo jack connector with Tip, Ring and Sleeve connections, often used to carry stereo signals, balanced mono signals or to get signals in and out of a console inert point.

TS Jack—Mono jack with Tip and Sleeve connections only. Used for unbalanced signal connections, such as guitar leads.

Tweeter—Type of loudspeaker specifically designed to reproduce high frequencies.

Unbalanced—A two-wire signal connection where a single inner signal conductor is surrounded by a tubular conductive screen, which also doubles as the return signal path. The screen helps reduce interference problems but it is not nearly as effective as a balanced interface.

Unison—To play or layer the same part using two or more different instruments (or voices).

USB—Short for Universal Serial Bus, a serial communications protocol used to connect many computer peripherals including audio interfaces, MIDI interfaces, external hard drives and keyboards. USB comes in USB1, USB2 and USB3 formats with progressively greater data rates.

Valve—Vacuum tube amplification device, known, also as a tube in the US.

Velcocity—The rate at which a key is depressed on a MIDI keyboard and used to control loudness.

Vibrato—Pitch modulation, which may be a playing technique used with stringed instrument or, in the case of electronic instruments and effects, controlled from an LFO.

VU Meter—A meter designed to display something approximating the average signal levels in approximately the same way as the human ear. 0VU normally denotes the nominal operation level of a device.

Warmth—Subjective term used to describe sound with enhanced low end or processed using subtle distortion and compression to make it sound a little larger than life.

Watt—Unit of electrical power.

Waveform—A visual representation of an electrical signal.

White Noise—A random signal where the energy distribution produces the same amount of noise power for each Hz of the spectrum. (cf. Pink Noise)

Word Clock—The accurate clock that regulates the A to D and D to A conversion processes and digital audio data transfers. Embedded information also identifies the start and end of each digital word or sample, and which samples belong to the left and right audio channels. While AES3 and S/PDIF embed clock signals within their data streams, it is often necessary (and sometimes beneficial) to connect a discrete word clock between equipment via their separate Word Clock In and Out connectors, usually BNC connectors.

Write—To save data to a digital storage medium or device. For example, mixer scenes or effects unit patches.

XLR—Type of connector commonly used to carry balanced audio signals from microphones. Various pin configurations are available though most microphones use the three-pin convention.

Xmax—The degree of linear excursion of which a loudspeaker driver is capable.

Y-Lead—A cable with a TRS jack at one end spitting out to two unbalanced TS jacks. Often used to connect patchbays or outboard equipment to console insert points.

Zero Crossing Point—The place where an analogue waveform crosses from being positive to negative or vice versa.

Zipper Noise—Audible steps heard when a parameter is being changed in a digital audio processor that doesn't include sophisticated data smoothing algorithms (such as linear interpolation) required to avoid it.

Index